The Story of the Nations

THE
STORY OF THE MOORS, IN SPAIN

The Story of the Nations

THE

STORY OF THE MOORS, IN SPAIN

BY

STANLEY LANE-POOLE, B. A ., M.R.A.S.

WITH THE COLLABORATION OF

ARTHUR GILMAN ,M.A

AUTHOR OF" A HISTORY OF THE AMERICAN PEOEPLE""THE STORY OF ROME , " "THE STORY OF THE SARACENS," ETC. ETC.

KALPAZ

ISBN: 978-93-5128-738-4(PB)

First Published in 1898
Indian Reprint in 2017

Published by

Kalpaz Publications
C-30, Satyawati Nagar,
Delhi – 110052
E-mail: kalpaz@hotmail.com
Website: kalpazpublications.com
Ph.: +91-9212142040

THE ALPUXARRAS.

PREFACE.

THE history of Spain offers us a melancholy contrast. Twelve hundred years ago, Tarik the Moor added the land of the Visigoths to the long catalogue of kingdoms subdued by the Moslems. For nearly eight centuries, under her Mohammedan rulers, Spain set to all Europe a shining example of a civilized and enlightened State. Her fertile provinces, rendered doubly prolific by the industry and engineering skill of her conquerors, bore fruit an hundredfold. Cities innumerable sprang up in the rich valleys of the Guadelquivir and the Guadiana, whose names, and names only, still commemorate the vanished glories of their past. Art, literature, and science prospered, as they then prospered nowhere else in Europe. Students flocked from France and Germany and England to drink from the fountain of learning which flowed only in the cities of the Moors. The surgeons and doctors of Andalusia were in the van of science: women were encouraged to devote themselves to serious study, and the lady doctor was not unknown among the people of Cordova. Mathematics, astronomy and botany, history, philosophy and jurisprudence were to be mastered in Spain, and Spain alone. The practical work of the field, the scientific methods of irrigation, the arts of fortification and shipbuilding, the highest and most elaborate products of the loom, the graver and the hammer, the potter's wheel and the mason's trowel, were brought to perfection by the Spanish Moors. In the practice of war no less than in the arts of peace they long stood supreme. Their fleets disputed the command of the Mediterranean with the Fatimites, while their armies carried fire and sword through the Christian marches. The Cid himself, the national hero, long fought on the Moorish side, and in all save education was more than half a Moor. Whatsoever makes a kingdom great and prosperous, whatsoever tends to refinement and civilization, was found in Moslem Spain.

In 1492 the last bulwark of the Moors gave way before the crusade of Ferdinand and Isabella, and with Granada fell all Spain's greatness. For a brief while, indeed, the reflection of the Moorish splendour cast a borrowed light upon the history of the land which it had once warmed with its sunny radiance. The great epoch of Isabella, Charles V., and Philip II., of Columbus, Cortes, and Pizarro, shed a last halo about the dying moments of a mighty State. Then followed the abomination of desolation, the rule of the Inquisition, and the blackness of darkness in which Spain has been plunged ever since. In the land where science was once supreme, the Spanish doctors became noted for nothing but their ignorance and incapacity, and the discoveries of Newton and

Harvey were condemned as pernicious to the faith. Where once seventy public libraries had fed the minds of scholars, and half a million books had been gathered together at Cordova for the benefit of the world, such indifference to learning afterwards prevailed, that the new capital, Madrid, possessed no public library in the eighteenth century, and even the manuscripts of the Escurial were denied in our own days to the first scholarly historian of the Moors, though himself a Spaniard. The sixteen thousand looms of Seville soon dwindled to a fifth of their ancient number; the arts and industries of Toledo and Almeria faded into insignificance; the very baths —public buildings of equal ornament and use— were destroyed because cleanliness savoured too strongly of rank infidelity. The land, deprived of the skilful irrigation of the Moors, grew impoverished and neglected; the richest and most fertile valleys languished and were deserted; most of the populous cities which had filled every district of Andalusia fell into ruinous decay; and beggars, friars, and bandits took the place of scholars, merchants, and knights. So low fell Spain when she had driven away the Moors. Such is the melancholy contrast offered by her history.

Happily we have here only to do with the first of these contrasted periods, with Spain in her glory under the Moors, not with Spain in her degradation under the Bourbons. We have endeavoured to present the most salient points in the eight centuries of Mohammedan rule without prejudice or extenuation, and while not neglecting the heroic characters and legends which appeal to the imagination of the reader, we have especially sought to give a clear picture of the struggle between races and creeds which formed the leading cause of political movement in mediæval Spain. The student who wishes to pursue the subject further than it has been possible to carry it in the limits of this volume should read the following authorities, to which we are deeply indebted. The most important is the late Professor Dozy's *Histoire des Musulmans d'Espagne* (4 vols., Leyden, 1861), and the same scholar's *Récherches sur l'histoire et la littérature de l'Espagne pendant le moyen âge* (2 vols., 3rd ed., Paris and Leyden, 1881). These works are full of valuable information presented in a form which, though somewhat fragmentary, is equally pleasing to the literary and the historical sense. Professor Dozy was an historian as well as an Orientalist, and his volumes are at once judicious and profound. Very useful, too, is Don Pasqual de Gayangos's translation of El-Makkary's *History of the Mohammedan Dynasties in Spain* (2 vols., London, 1843), which has been exposed to some needlessly acrimonious criticism by Professor Dozy and others on the score of certain minor inaccuracies, but which none the less deserves the gratitude of all students who would rather have half a loaf than no bread, and are glad to be able to read an Arabic writer, even imperfectly, in a European tongue. Don Pasqual's notes, moreover, present a mass of valuable material which can be obtained nowhere else. Beyond these two authorities there are many Arabic historians, whose works have been consulted in the composition of the present volume, but who can hardly be recommended to the general student, as very few of them have found translators. A slight but very readable and instructive sketch of Arab civilization, with a glance at the Spanish development, is found in August Bebel's *Die Mohammedanisch-arabische Kulturperiode* (Stuttgart, 1884). For the last days of the Moorish domination, Washington Irving's picturesque *Conquest of Granada*, and Sir W. Stirling Maxwell's admirable *Don John of Austria*, largely drawn upon in this volume, deserve separate reading. All histories of the Moors written before the works of Gayangos and Dozy should be studiously avoided, since they are mainly founded upon Conde's *Dominacion de los Arabes in España*, a book of considerable literary merit but very slight historical value, and the source of most of the errors that are found in later works. Whether it has been in any degree the foundation of Miss Yonge's *Christians and Moors in Spain* (the only popular history of this period in English of which I have heard), I cannot determine: for a glance at her pages, while exciting my admiration, showed me that her book was written so much on the lines which I had drawn for my own work that I could not read it without risk of involuntary imitation.

Besides my indebtedness to the works of Dozy and Gayangos, and to the kind collaboration of Mr. Arthur Gilman, I have gratefully to acknowledge the assistance of my friend Mr. H. E. Watts, especially in matters of Spanish orthography.

In conclusion, those who are inclined to infer, from the picture here given of Moorish civilization, that

Mohammedanism is always on the side of culture and humanity, must turn to another volume in this series, my *Story of the Turks*, to see what Mohammedan barbarism means. The fall of Granada happened within forty years of the conquest of Constantinople; but the gain to Islam in the east made no amends for the loss to Europe in the west: the Turks were incapable of founding a second Cordova.

S. L.-P.

RICHMOND. SURREY,
July, 1886.

CONTENTS.

PAGE

I.

The Last of the Goths 1

The seclusion of Ancient Arabia, 1—Change caused by the Prophet Mohammed, 2—The Saracen conquests, 3—Ceuta attacked, 4—Condition of Spain, 4—Effects of Roman rule, 5—The Visigoths, 6—Demoralization of all classes, 7—Witiza, 8—Roderick, 8—Story of Florinda, 11—Count Julian's revenge, 11—He joins the Arabs, 12—Mūsa son of Noseyr, 12—First incursion into Spain under Tarif, 13—Tārik's invasion, 13—The Enchanted Tower, 14—Roderick's vision, 18—Battle of the Guadalete, 20—Fate of Don Rodrigo, 21.

II.

The Wave of Conquest 23

Subjugation of Spain, 23—Capture of Cordova, Malaga, Elvira, Murcia, 24—Theodemir's stratagem, 25—Flight of the Goths, 26—Mūsa crosses over to Spain, 27—His jealousy of Tārik, and recall, 28—Invasion of Aquitaine, and capture of Narbonne, 28—Battle of Tours, 29—A boundary set to the Moorish advance by Charles Martel, 30—Charlemagne invades Spain, 33—The Pass of Roncesvalles, 34—Death of Roland, 36.

PAGE

III.

THE PEOPLE OF ANDALUSIA 39

The limits of the Moorish territory, 39—Division between the north and the south, 40—Andalusia, 43—Condition of the people after the Conquest, 44—Taxation, 47—Moderation of the Moors, 47—State of the slaves, 48—The renegades, 49—Factions among the victors, 50—Arab tribal jealousies, 51—The Berbers or Moors proper, 52—Their superstitious character, 53—Berber insurrections in Africa and Spain, 54—Syrian Arabs come to the rescue, 55—Their settlement in Andalusia, 56.

IV.

A YOUNG PRETENDER 58

The Khalifs of Damascus, 58—Overthrow of the Omeyyads, 59—Adventures of Abd-er-Rahmān the Omeyyad, 60—He lands in Spain and is received with acclamation, 62—Foundation of the Omeyyad kingdom of Andalusia, 63—Revolts suppressed by Abd-er-Rahmān, 64—His character, 66—Hishām I., 71—His piety and virtues, 71—Power of the priests, 72—Yahya the theologian, 73—Accession of Hakam 74—His genial character, 74—Revolt of the zealots, 75—Burning of the southern suburb of Cordova, 76.

V.

THE CHRISTIAN MARTYRS 78

Abd-er-Rahmān II., 78—Queen Tarūb, 81—Ziryāb the exquisite, 81—Frivolity of the Court, 82—Christian fanaticism, 84—A race for martyrdom, 85—St. Eulogius and Flora, 86—Death of Perfectus, 89—More "martyrs," 90—Indifference of the majority of the Christians, 90—Moderation counselled by the Church, 91—Flora and Eulogius in prison, 92—Their martyrdom, 93.

VI.

THE GREAT KHALIF 96

Large movements of race and creed in Andalusia, 96—The need of a great king, 98—Abdallah's weakness, 98—General anarchy, 101—Ibn-Hafsūn's rebellion, 102—Ibn-Hajjāj of

PAGE

Seville, 105—Cordova in danger, 106—Accession of Abd-er-Rahmān III., 107—His courageous policy, 108—Submission of the rebels, 109—Death of Ibn-Hafsūn and conquest of Bobastro, 110—Siege of Toledo, 110—Surrender, 113—Pacification of Andalusia, 113.

VII.

THE HOLY WAR 114

Abd-er-Rahmān's principle of government, 114—The Slavs, 114—Wars with the Fātimite Khalifs of Africa, 115—Pelayo and the Christians of the Asturias, 116—Growth of the Christian power, 117—Alfonso's campaigns, 118—The soldiery of Leon, 119—Ordoño's forays, 119—Battle of St. Estevan de Gormaz, 120—Abd-er-Rahmān retaliates, 120—Battle of the Val de Junqueras and capture of Pamplona, 121—Abd-er-Rahmān assumes the title of Khalif, 121—Annual campaigns against the Christians, 122—Ramiro defeats him at Alhandega, 123—Jealousies among the Christians, 123—Fernando Gonzalez, 123—Queen Theuda and Sancho the Fat invoke the Khalif's aid, 125—Their visit to Cordova, 126—Hazdai the physician, 126—Death of Abd-er-Rahmān III., 126—His achievements and character, 127.

VIII.

THE CITY OF THE KHALIF 129

Beauty of Cordova, 129—Gardens, 131—Palaces, 132—Baths, 135—The Great Mosque, 136—"The City of the Fairest," 139—Reception at Medinat-ez-Zahrā, 142—Science and letters cultivated under the Moors, 144—Condition of the arts in Andalusia, 147.

IX.

THE PRIME MINISTER 152

Hakam II., 152—His library, 155—Hishām II., 156—Seclusion in the harīm, 156—The Queen-mother Aurora, 156—Harīm influence, 157—Rise of Ibn-Abī-Amir, surnamed Almanzor, 157—His campaign with Ghālib against the Christians, 159—He becomes Prime Minister, 160—His absolute

THE ALPUXARRAS.

rule, 161—Policy, 162—Fortitude, 162—Resource, 162—The new army, 163—Campaigns against the Christians of the North, 164—Invasion of Leon, Barcelona, and Galicia, 165—Capture of St. Santiago de Compostella, 165—Unchecked victories, 166—Death, 166—"Buried in Hell," 166.

X.

THE BERBERS IN POWER 167

Anarchy after Almanzor's death, 167—His sons, 169—Succession of puppet Khalifs, 170—Misery of Hishām III., 171—Massacres and pillaging, 172—The Slavs and the Berbers, 175—Sack of the City of Ez-Zahrā, 175—Petty dynasties, 176—Advance of the Christians of Leon and Castile, 176—Alfonso VI., 177—The Cid, 177—The Moors call in the Almoravides, 178—Battle of Zallāka, 179—Character of the Almoravides, 180—They subdue Andalusia, 181—Their tyranny and demoralization, 183—The expulsion of the Almoravides, 184.

XI.

MY CID THE CHALLENGER 185

State of the Christian powers in the North, 185—Fernando I., 186—Vassalage of the Mohammedan princes, 186—Character of the Christians and Moors contrasted, 189—The *chevaliers d'industrie*, 191—The Cid Rodrigo de Bivar, 191—His title of Campeador, 191—His panegyrists, 192—Dozy's "real Cid," 192—The *Chronicle of the Cid*, 193—Heroic character, 193—The Cid's first appearance in history, 195—His services to Castile, 195—His banishment, 195—Takes service with the Moorish king of Zaragoza, 200—Fights against the Christians of Barcelona, 201—At Valencia, 205—Raid upon Leon, 206—Siege of Valencia, 206—Battle with the Almoravides, 209—Death and burial of the Cid, 213.

XII.

THE KINGDOM OF GRANADA 214

Invasion of Andalusia by the Almohades, 214—Victory at Alarcos, 217—Defeat at Las Navas, 217—Expulsion of the

PAGE

Almohades, 217—Advance of the Christians, 217—Granada alone left to the Moors, 218—Dynasty of the Beny-Nasr of Granada, 218—Their tribute to Castile, 221—The Alhambra, 221—Ferdinand and Isabella, 232—Abul-Hasan (Alboacen) throws off his allegiance, 232—Capture of Zahara, 233—Fall of Alhama, 235—Disasters of the Christians in the mountains of Malaga, 236—Defeat of the Moors at Lucena, 242—Boabdil made prisoner, 245.

XIII.

THE FALL OF GRANADA 246

Ferdinand's policy towards Boabdil, 246—Factions at Granada, 247—The Abencerrages, 247—Ez-Zaghal, 248—Ferdinand's campaigns, 251—Siege of Velez and Malaga, 251—Ez-Zegry's defence, 253—The surrender, 254—Siege of Baza, 258—Ez-Zaghal submits, 259—His fate, 259—Granada threatened, 260—Mūsa's reply, 260—The siege, 263—Exploit of Pulgar, 264—Boabdil capitulates, 266—Death of Mūsa, 266—Entry of Ferdinand and Isabella into the Alhambra, 266—"The last sigh of the Moor," 267.

XIV.

BEARING THE CROSS 269

Terms of surrender of Granada, 269—Archbishop Talavera's toleration, 269—Cardinal Ximenes, 269—Revolt in the Alpuxarras, 271—Defeat and death of Aguilar, 271—Persecution of the Moriscos, 272—Second revolt in the Alpuxarras, 274—Character of the country, 274—Heroism of the Christians, 276—The plank of Tablete, 276—Massacre of the Moors in the Albaycin gaol, 277—Aben Umeyya and Aben Abó, 277—Don John of Austria, 278—Banishment of the Moors, 279—Rejoicings in Spain, 279—Retribution, 280.

INDEX 281

LIST OF ILLUSTRATIONS.

	PAGE
THE ALPUXARRAS	*Frontispiece*
TOLEDO	9
GATE OF BISAGRA, TOLEDO	15
PUERTO DEL SOL, TOLEDO	27
ARCH IN THE ALJAFERIA OF ZARAGOZA . . .	31
ALCANTARA	41
THE SIERRA NEVADA	45
THE BRIDGE OF CORDOVA	69
MOORISH IVORY CASKET OF THE 11TH CENTURY IN THE CATHEDRAL OF PAMPLONA	79
THE GOLDEN TOWER, SEVILLE	99
DOOR OF THE MAIDEN'S COURT, ALCAZAR OF SEVILLE	103
AQUEDUCT NEAR GRANADA	111
EXTERIOR OF THE GREAT MOSQUE AT CORDOVA .	133
GATE OF THE MOSQUE OF CORDOVA	137
HISPANO-MORESCO VASE. (*Preserved at Granada*) .	145
HISPANO-MORESCO LUSTRED PLATE, WITH ARMS OF LEON, CASTILE, AND ARAGON. (*In the South Kensington Museum*).	149
ANCIENT KORAN CASE. (*Escurial Library*) . .	153

PAGE

THE GIRALDA AT SEVILLE 173

BOTICA DE LOS TEMPLARIOS, TOLEDO 187

GATE OF SERRANO, VALENCIA 203

TOMB OF THE CID AT SAN PEDRO DE CARDEÑA . 211

BANNER OF THE ALMOHADES 215

SHIELD OF A KING OF GRANADA 219

THE COURT OF THE LIONS IN THE ALHAMBRA . . 223

GARDEN OF THE GENERALIFE, GRANADA . . . 229

A WINDOW IN THE ALHAMBRA 243

MOSQUE LAMP FROM GRANADA 249

MALAGA 255

SWORD OF BOABDIL (*Villaseca Collection, Madrid*) . 261

THE STORY OF THE MOORS IN SPAIN.

I.

THE LAST OF THE GOTHS.

WHEN the armies of Alexander the Great were trampling upon the ancient empires of the East, one country remained undisturbed and undismayed. The people of Arabia sent no humble embassies to the conqueror. Alexander resolved to bring the contemptuous Arabs to his feet: he was preparing to invade their land when death laid its hand upon him, and the Arabs remained unconquered.

This was more than three hundred years before Christ, and even then the Arabs had long been established in independence in their great desert peninsula. For nearly a thousand years more they continued to dwell there in a strange solitude. Great empires sprang up all around them; the successors of Alexander founded the Syrian kingdom of the Seleucids and the Egyptian dynasty of the Ptolemies; Augustus was crowned Imperator at Rome; Constantine became the first Christian emperor at Byzantium;

the hordes of the barbarians bore down upon the wide-reaching provinces of the Cæsars—and still the Arabs remained undisturbed, unexplored, and unsubdued. Their frontier cities might pay homage to Chosroes or Cæsar, the legions of Rome might once and again flash across their highland wastes; but such impress was faint and transitory, and left the Arabs unmoved. Hemmed in as they were by lands ruled by historic dynasties, their deserts and their valour ever kept out the invader, and from the days of remote antiquity to the seventh century of the Christian era hardly anything was known of this secluded people save that they existed, and that no one attacked them with impunity.

Then suddenly a change came over the character of the Arabs. No longer courting seclusion, they came forth before the world, and proceeded in good earnest to conquer it. The change had been caused by one man. Mohammed the Arabian Prophet began to preach the religion of *Islam* in the beginning of the seventh century, and his doctrine, falling upon a people prone to quick impulses and susceptible of strong impressions, worked a revolution. What he taught was simple enough. He took the old faith of the Hebrews, which had its disciples in Arabia, and, making such additions and alterations as he thought needful, he preached the worship of One God as a new revelation to a nation of idolaters. It is difficult for us in the present time to understand the irresistible impulse which the simple and unemotional creed of Mohammed gave to the whole people of Arabia; but we know that such religious revolutions have been, and that there is always a mysterious and potent fascination in the personal influence of a true prophet. Mohammed was so far true, that he taught honestly and strenuously what he believed to be the only right faith, and there was enough of sublimity in the creed and of enthusiasm in the Prophet and his hearers to produce that wave of overmastering popular feeling which people call fanaticism. The Arabs before the time of Mohammed had been a collection of rival tribes or clans, excelling in the savage virtues of bravery, hospitality, and even chivalry, and devoted to the pursuit of booty. The Prophet turned the Arab tribes, for the nonce, into the Moslem people, filled them with the fervour of martyrs, and added to the greed of plunder the nobler ambition of bringing all mankind to the knowledge of the truth.

Before Mohammed died he was master of Arabia, and the united tribes who had embraced the Moslem or Mohammedan faith were already spreading over the neighbouring lands and subduing the astonished nations. Under his successors the Khalifs, the armies of the Mussulmans overran Persia and Egypt and North Africa as far as the Pillars of Hercules; and the Muezzins chanted the Call to Prayer to the Faithful over all the land from the river Oxus in Central Asia to the shores of the Atlantic Ocean.

The Mohammedans, or Saracens (a word which means "Easterns"), were checked in Asia Minor by the forces of the Greek Emperor; and it was not till the fifteenth century that they at last obtained the long-coveted possession of Constantinople, by the valour of the Ottoman Turks. So, too, at the opposite extremity of the Mediterranean, it was an officer of the Greek Emperor who for a while held the Arab advance in check. The conquerors swept over the provinces of North Africa, and, after a long struggle, reduced the turbulent Berber tribes for a while to submission, till only the fortress of Ceuta held out against them. Like the rest of the southern shore of the Mediterranean, Ceuta belonged to the Greek Emperor; but it was so far removed from Constantinople that it was thrown upon the neighbouring kingdom of Spain for support, and, while still nominally under the authority of the Emperor, looked really to the King of Toledo for assistance and protection. It is not likely that all the aid that Spain could have given would have availed against the surging tide of Saracen invasion; but, as it happened, there was a quarrel at that time between Julian the governor of Ceuta and Roderick the King of Spain, which opened the door to the Arabs.

Spain was then under the rule of the Visigoths, or West Goths, a tribe of barbarians, like the many others who overran the provinces of the Roman Empire in its decline. The Ostrogoths had occupied Italy; and their kinsmen the Visigoths, displacing or subduing the Suevi (or Swabians) and other rude German tribes, established themselves in the Roman province of Iberia (Spain) in the fifth century after Christ. They found the country in the same condition of effeminate luxury and degeneracy that had proved the ruin of other parts of the empire. Like many warlike peoples, the Romans, when their work was accomplished and the world was at their feet, had rested contentedly from their labours, and abandoned themselves to the pleasures that wealth and security permit. They were no longer the brave stern men who lived simple lives and left the ploughshare to wield the sword when a Scipio or a Cæsar summoned them to defend their country or to conquer a continent. In Spain the richer classes were given over to luxury and sensuality; they lived only for eating and drinking, gambling and all kinds of excitement. The mass of the people were either slaves, or, what was much the same thing, labourers bound to the soil, who could not be detached from the land they cultivated but passed with it from master to master. Between the rich and the slaves was a middle class of burghers, who were perhaps even worse off: for on their shoulders lay all the burden of supporting the State; they paid the taxes, performed the civil and municipal functions, and supplied the money which the rich squandered upon their luxuries. In a society so demoralized there were no elements of opposition to a resolute invader. The wealthy nobles were too deeply absorbed in their pleasures to be easily roused by rumours of an enemy; their swords were rusty with being too long laid aside. The slaves felt little interest in a change of masters, which could hardly make them more miserable than they already were; and the burghers were discontented with the arrangement of the burdens of the State, by which they had to bear most of the cost while they reaped none of the advantages.

Out of such men as these a strong and resolute army could not be formed; and the Goths therefore entered Spain with little trouble ; the cities willingly opened their gates, and the diseased civilization of Roman Spain yielded with hardly a blow. The truth was that the road of the Goths had been too well prepared by previous hordes of barbarians—Alans, Vandals, and Suevi — to need much exertion on their own part. The Romanized Spaniards had fully learned what a barbarian invasion entailed : they had seen their cities burnt, their wives and children carried captives, those few leaders who showed any manly resistance massacred ; they had seen the consequences of the barbarian scourge—plague and famine, wasted lands, starving inhabitants, and everywhere savage anarchy. They had learned their lesson, and meekly admitted the Goths.

In the beginning of the eighth century, when the Saracens had reached the African shore of the Atlantic and were looking across the Straits of Hercules to the sunny provinces of Andalusia, the Goths had been in possession of Spain for more than two hundred years. There had been time enough to reform the corrupt condition of the kingdom and to infuse the fresh vigour of youth which an old civilization sometimes gains by the introduction of barbarous but masculine races. There were special reasons why the Goths should improve the state of Spain. They were not only bold, strong, and uncorrupted by ease of life ; they were Christians, and, in their way, very earnest Christians. Spain was but nominally converted at the time of their arrival : Constantine had indeed promulgated Christianity as the religion of the Roman Empire, but it had taken very little root in the Western provinces. The advent of an ignorant but devout race like the Goths might probably arouse a more earnest faith in the new religion amid the worn-out paganism of the kingdom, and the Catholic priests were full of hope for the future of their church. The result did not in any way justify the anticipation. The Goths remained devout indeed, but they regarded their acts of religion chiefly as reparation for their vices; they compounded for exceptionally bad sins by an added amount of repentance, and then they sinned again without compunction. They were quite as corrupt and immoral as the Roman nobles who had preceded them, and their style of Christianity did not lead them to endeavour to improve the condition of their subjects. The serfs were in an even more pitiable state than before. Not only were they tied to the land or master, but they could not marry without his consent, and if slaves of neighbouring estates intermarried, their children were distributed between the owners of the several properties. The middle classes bore, as in Roman times, the burden of taxation, and were consequently bankrupt and ruined: the land was still in the hands of the few, and the large estates were indifferently cultivated by crowds of miserable slaves, whose dreary lives were brightened by no hope of improvement or dream of release before death. The very clergy, who preached about the brotherhood of Christians, now that they had become rich and owned great estates, joined in the traditional policy and treated their slaves and serfs as badly as any Roman noble. The rich were sunk in the same slough of sensuality that had proved the ruin of the Romans, and the vices of the Christian Goths rivalled, if they did not exceed, the polished wickedness of the pagans. "King Witiza," says the chronicler, anxious to find some reason for the overthrow of the Christians by the Saracens, "taught all Spain to sin." Spain, indeed, knew only too well how to sin before, and Witiza may have been no worse than his predecessors; but the Goths gave a fresh license to the general corruption. The vices of barbarians show often a close resemblance to those of decayed civilization, and in this instance the change of rulers brought no amelioration of morals.[1]

Such was the condition of Spain when the Mussulman approached her borders. A corrupt aristocracy divided the land among themselves; the great estates were tilled by a wretched and hopeless race of serfs; the citizen classes were ruined. On the other side of the straits of Gibraltar were the soldiers of Islam, all hardy warriors, fired with the fervour of a new faith, bred to arms from their childhood, simple and rude in their life, and eager to plunder the rich lands of the infidels. Between two such peoples there could be no doubt as to the issue of the fight; but to remove the possibility of doubt, treachery came to the aid of the invaders.

Witiza had been deposed by Roderick, a prince who seems to have begun his reign well, but who presently succumbed to the temptations of wealth and power. His selfish pleasure-loving disposition set fire to the combustible materials that sur-

[1] Dozy: Hist. des Mus. d'Espagne, livre ii. ch. i.

TOLEDO.

rounded him and that needed but a spark to explode and destroy his kingdom. It was then the custom among the princes of the State to send their children to the court, to be trained in whatever appertained to good breeding and polite conduct. Among others, Count Julian, the governor of Ceuta, sent his daughter Florinda to Roderick's court at Toledo to be educated among the queen's waiting women. The maiden was very beautiful, and the king, forgetful of his honour, which bound him to protect her as he would his own daughter, put her to shame.[1] The dishonour was the greater, since Julian's wife was a daughter of Witiza, and the royal blood of the Goths had thus been insulted in the person of Florinda. In her distress the young girl wrote to her father, and, summoning a trusty page, bade him, if he hoped for knightly honour or lady's favour, to speed with all haste, night and day, over land and sea, till he placed the letter in Count Julian's hand.

Julian had no reason to love King Roderick; his own connection with the deposed and probably murdered King Witiza forbade fellowship with the usurper; and his daughter's dishonour fanned his smouldering rancour to a blaze of vengeful fury. He had so far successfully resisted the attacks of the Arabs; but now he resolved no longer to defend the kingdom of his daughter's destroyer. The Saracens should have Spain if they would, and

[1] I reproduce this celebrated legend without vouching for its truth. Florinda, or Cava as the Moslems call her, plays too prominent a part in the first chapter of Andalusian history to be ignored; and, if her part be fictitious, her father's treachery at least is certain.

he was ready to show them the way. Full of a passion for revenge, Julian hastened to the Court of Roderick, where he so skilfully disguised his mind that the king, who felt some remorse and trusted that Florinda had kept the secret, heaped honours upon him, took his counsel in everything relating to the defence of the kingdom, and even by his treacherous advice sent the best horses and arms in Spain to the south under Julian's command, to be ready against the infidel invaders. Count Julian departed from Toledo in the highest favour of the king, taking his daughter with him. Roderick's parting request was that the Count would send him some special kind of hawks, which he needed for hunting; Julian made answer, that he would bring him such hawks as he had never in his life seen before, and with this covert hint of the coming of the Arabs he went back to Ceuta.

As soon as he had returned, he paid a visit to Mūsa, the son of Noseyr, the Arab governor of North Africa, with whom his troops had many times crossed swords, and he told him that war was now over between them—henceforth they must be friends. Then he filled the ears of the Arab general with stories of the beauty and richness of Spain, of its rivers and pastures, vines and olives, its splendid cities and palaces, and the treasures of the Goths: it was a land flowing with milk and honey, he said, and Mūsa had only to go over and take it. Julian himself would show him the way, and lend him the ships. The Arab was a cautious general, however; this inviting proposal, he considered, might cover a treacherous ambuscade; so he sent messengers to his master the

Khalif at Damascus, to ask for instructions, and meantime contented himself with sending a small body of five hundred men, under Tarīf, in 710, to make a raid, in Julian's four ships, upon the coast of Andalusia. The Arabs had not yet become used to the navigation of the Mediterranean, and Mūsa was unwilling to expose more than an insignificant part of his army to the perils of the deep.

Tarīf returned in July, having successfully accomplished his mission. He had landed at the place which still bears his name, Tarīfa, had plundered Algeciras, and seen enough to assure him that Count Julian's tale of the defenceless state of Spain was true, and that his own loyalty to the invaders was to be depended upon. Still Mūsa was not disposed to venture much upon the new conquest. The Khalif of Damascus had enjoined him on no account to risk the whole Moslem army in unknown dangers, and had only authorized small foraying expeditions. Still, encouraged by Tarīf's success, Mūsa resolved upon a somewhat larger venture. In 711, learning that Roderick was busy in the north of his dominions, where there was a rising of the Basques, Mūsa despatched one of his generals, the Moor Tārik, with 7,000 troops, most of whom were also Moors,[1] to make another raid upon Andalusia. The raid carried him further than he expected. Tārik landed at the lion's rock, which has ever since borne his name.

[1] The word Moor is conveniently used to signify Arabs and other Mohammedans in Spain, but properly it should only be applied to *Berbers* of North Africa and Spain. In this volume the term is used in its common acceptation, unless the Arabs are specially distinguished from the Berbers.

Gebal-Tarik, Gibraltar, and after capturing Carteya, advanced inland. He had not proceeded far when he perceived the whole force of the Goths under Roderick advancing to encounter him. The two armies met on the banks of a little river, called by the Saracens the Wady Bekka, near the Guadalete, which runs into the Straits by Cape Trafalgar.

The legend runs that some time before this, as King Roderick was seated on his throne in the ancient city of Toledo, two old men entered the audience chamber. They were arrayed in white robes of ancient make, and their girdles were adorned with the signs of the Zodiac and hung with innumerable keys. "Know, O king," said they, "that in days of yore, when Hercules had set up his pillars at the ocean strait, he erected a strong tower near to this ancient city of Toledo, and shut up within it a magical spell, secured by a ponderous iron gate with locks of steel; and he ordained that every new king should set a fresh lock to the portal, and foretold woe and destruction to him who should seek to unravel the mystery of the tower. Now, we and our ancestors have kept the door of the tower from the days of Hercules even to this hour; and though there have been kings who have sought to discover the secret, their end has ever been death or sore amazement. None ever penetrated beyond the threshold. Now, O king, we come to beg thee to affix thy lock upon the enchanted tower, as all the kings before thee have done." Whereupon the aged men departed.

But Roderick, when he had thought of all they had said, became filled with a burning desire to enter the

GATE OF BISAGRA, TOLEDO.

enchanted tower, and despite the warnings of his bishops and counsellors, who told him again that none had ever entered the tower alive, and that even great Cæsar had not dared to attempt the entrance—

> Nor shall it ever ope, old records say,
> Save to a king, the last of all his line,
> What time his empire totters to decay,
> And treason digs, beneath, her fatal mine,
> And high above, impends avenging wrath Divine—

despite all admonition, he rode forth one day, accompanied by his cavaliers, and approached the tower. It stood upon a lofty rock, and cliffs and precipices hemmed it in. Its walls were of jasper and marble, inlaid in subtle devices, which shone in the rays of the sun. The entrance was through a passage cut in the stone, and was closed by the great iron gate covered with the rusty locks of all the centuries from the time of Hercules to Witiza; and on either hand stood the aged men who had come to the audience hall. All day long did the two old janitors, though foreboding ill, aided by Roderick's gay cavaliers, labour to turn the rusty keys, until, when it was near sundown, the gate was undone, and the king and his train advanced to the entrance. The gate swung back, and they entered a hall, on the other side of which, guarding a second door, stood a gigantic bronze figure of terrible aspect, which wielded a huge mace unceasingly and dealt mighty blows upon the earth around.

When Roderick saw this figure, he was dismayed awhile; but seeing on its breast the words, "I do my duty," he plucked up courage and conjured it to let him pass in safety, for he meant no sacrilege, but only wished to learn the mystery of the tower. Then the figure stood still, with its mace uplifted, and the king and his followers passed beneath it into the second chamber. They found this encrusted with precious stones, and in its midst was a table, set there by Hercules, and on it a casket, with the inscription, "In this coffer is the mystery of the Tower. The hand of none but a king can open it; but let him beware, for wonderful things will be disclosed to him, which must happen before his death."

When the king had opened the coffer, there was nothing in it but a parchment folded between two plates of copper; on it were figured men on horseback, fierce of countenance, armed with bows and scimitars, and above them was the motto, "Behold, rash man, those who shall hurl thee from thy throne and subdue thy kingdom." And as they gazed upon the picture, on a sudden they heard the sound of warfare, and saw, as though in a cloud, that the figures of the strange horsemen began to move, and the picture became a vision of war:

> So to sad Roderick's eye, in order spread,
> Successive pageants filled that mystic scene,
> Showing the fate of battles ere they bled,
> And issue of events that had not been.

"They beheld before them a great field of battle, where Christians and Moors were engaged in deadly conflict. They heard the rush and tramp of steeds, the blast of trump and clarion, the clash of cymbal, and the stormy din of a thousand drums. There was the flash of swords and maces and battle-axes, with the whistling of arrows and the hurling of darts and lances. The Christians quailed before the foe. The infidels pressed upon them and put them to utter rout; the standard of the Cross was cast down, the banner of Spain was trodden under foot; the air resounded with shouts of triumph, with yells of fury, and with the groans of dying men. Amidst the flying squadrons King Roderick beheld a crowned warrior, whose back was turned towards him, but whose armour and device were his own, and who was mounted on a white steed that resembled his own war-horse Orelia. In the confusion of the fight, the warrior was dismounted, and was no longer seen to be, and Orelia galloped wildly through the field of battle without a rider." [1]

When the king and his attendants fled dismayed from the enchanted tower, the great bronze figure had disappeared, the two aged janitors lay dead at the entrance, and amid various stormy portents of nature the tower burst into a blaze, and every stone was consumed and scattered to the winds; and it is related that wherever its ashes fell to the earth there was seen a drop of blood.

The mediæval chroniclers, both Christian and Arab, delighted to relate portents such as these:

> Legend and vision, prophecy and sign,
> Where wonders wild of Arabesque combine
> With Gothic imagery of darker shade;

and we read how both sides of the approaching combat were cheered or dismayed by omens of various

[1] Washington Irving: The Conquest of Spain, Bohn's ed., 378 ff.; American edition, Spanish Papers, vol. i. p. 42.

kinds. The Prophet himself is said to have appeared to Tārik, and to have bidden him be of good courage, to strike, and to conquer; and many like fables are related. But whatever may have been the dreams and visions of the armies then encamped over against one another near the river Guadelete, the result of the combat was never doubtful. Tārik, indeed, although he had been reinforced with 5,000 Berbers, commanded still but a little army of 12,000 troops, and Roderick had six times as many men to his back. But the invaders were bold and hardy men, used to war, and led by a hero; the Spaniards were a crowd of ill-treated slaves, and among their commanders were treacherous nobles. The kinsmen of Witiza were there, obedient to the summons of Roderick; but they intended to desert to the enemy's side in the midst of the battle and win the day for the Saracens. They had no idea that they were betraying Spain. They thought that the invaders were only in search of booty; and that, the raid over and the booty secured, they would go back to Africa, when the line of Witiza would be restored to its ancient seat. And thus they lent a hand to the day's work which placed the fairest provinces of Spain for eight centuries under the Moslem domination.

When the Moors saw the mighty army that Roderick had brought against them, and beheld the king in his splendid armour under a magnificent canopy, their hearts for a moment sank within them. But Tārik cried aloud, "Men, before you is the enemy, and the sea is at your backs. By Allah, there is no escape for you save in valour and resolution." And they plucked up courage and shouted, "We will follow thee, O Tārik," and rushed after their general into the fray. The battle lasted a whole week, and prodigies of valour are recorded on both sides. Roderick rallied his army again and again; but the desertion of the partisans of Witiza turned the fortune of the field and it became the scene of a disastrous rout.

The hosts of Don Rodrigo were scattered in dismay,
When lost was the eighth battle, nor heart nor hope had they;
He, when he saw that field was lost, and all his hope was flown,
He turned him from his flying host, and took his way alone.

All stained and strewed with dust and blood, like to some smouldering
brand
Plucked from the flame, Rodrigo showed: his sword was in his hand,
But it was hacked into a saw of dark and purple tint:
His jewelled mail had many a flaw, his helmet many a dint.

He climbed into a hill-top, the highest he could see,
Thence all about of that wide rout his last long look took he;
He saw his royal banners, where they lay drenched and torn,
He heard the cry of victory, the Arab's shout of scorn.

He looked for the brave captains that led the hosts of Spain,
But all were fled except the dead, and who could count the slain?
Where'er his eye could wander, all bloody was the plain,
And while thus he said, the tears he shed ran down his cheeks like rain:

"Last night I was the King of Spain—to-day no king am I;
Last night fair castles held my train—to-night where shall I lie?
Last night a hundred pages did serve me on the knee—
To-night not one I call my own—not one pertains to me.

O luckless, luckless was the hour, and cursed was the day,
When I was born to have the power of this great seniory!
Unhappy me, that I should see the sun go down to-night!
O Death, why now so slow art thou, why fearest thou to smite?[1]

So runs the old Spanish ballad; but the fate of Roderick has remained a mystery to this day. His

[1] Lockhart: Spanish Ballads.

horse and sandals were found on the river bank the day after the battle ; but his body was not with them. Doubtless he was drowned and washed out to the great ocean. But the Spaniards would not believe this. They clothed the dead king with a holy mystery which assuredly did not enfold him when alive. They made the last of the Goths into a legendary saviour like King Arthur, and believed that he would come again from his resting-place in some ocean isle, healed of his wound, to lead the Christians once more against the infidels. In the Spanish legends, Roderick spent the rest of his life in pious acts of penance, and was slowly devoured by snakes in punishment for the sins he had committed, until at last his crime was washed out, " the body's pang had spared the spirit's pain," and " Don Rodrigo " was suffered to depart to the peaceful isle, whence his countrymen long awaited his triumphant return.

II.

THE WAVE OF CONQUEST.

"O COMMANDER of the Faithful, these are not common conquests; they are like the meeting of the nations on the Day of Judgment." Thus wrote Mūsa, the Governor of Africa, to the Khalif Welīd, describing the victory of the Guadalete. There is little wonder that the Saracens stood amazed at the completeness of their triumph. Leaving the regions of myth, with which the Spanish chroniclers have surrounded the fall of Roderick, it is matter of sober history that the victory of the Guadalete gave all Spain into the hands of the Moors. Tārik and his twelve thousand Berbers had by a single action won the whole peninsula, and it needed but ordinary energy and promptness to reduce the feeble resistance which some of the cities still offered. The victor lost no time in following up his success. In defiance of an order from Mūsa, who was bitterly jealous of the unexpected glory which had come to his Berber lieutenant, and commanded him to advance no further, the fortunate general pushed on without delay. Dividing his forces into three brigades, he spread them over the peninsula, and reduced city after city with little difficulty. Mughīth, one of his officers, was despatched with seven hundred horse to seize Cordova. Lying hid till darkness came on, Mughīth stealthily approached the city. A storm of hail, which the Moslems regarded as a special favour of Providence, muffled the clatter of their horses' hoofs. A shepherd pointed out a breach in the walls, and here the Moors determined to make the assault. One of them, more active than the rest, climbed a fig-tree which grew beneath the breach, and thence, springing on to the wall, flung the end of a long turban to the others, and pulled them up after him. They instantly surprised the guard, and threw open the gates to the main body of the invaders, and the town was captured with hardly a blow. The governor and garrison took refuge in a convent, where for three months they were closely beleaguered. When at length they surrendered, Cordova was left in the keeping of the Jews, who had proved themselves staunch allies of the Moslems in the campaign, and who ever afterwards enjoyed great consideration at the hands of the conquerors. The Moors admitted them to their intimacy, and, until very late times, never persecuted them as the Gothic priests had done. Wherever the arms of the Saracens penetrated, there we shall always find the Jews in close pursuit: while the Arab fought, the Jew trafficked, and when the fighting was over, Jew and Moor and Persian joined in that cultivation of learning and philosophy, arts and sciences, which preëminently distinguished the rule of the Saracens in the Middle Ages.

With the coöperation of the Jews, and the terror of the Spaniards, Tārik's conquest proceeded apace.

Archidona was occupied without a struggle: the inhabitants had all fled to the hills. Malaga surrendered, and Elvira (near where Granada now stands) was stormed. The mountain passes of Murcia were defended by Theodemir for some time with great valour and prudence; but at last, being over-persuaded into offering a pitched battle on the plain, the Christian army was cut to pieces, and Theodemir escaped with a single page to the city of Orihuela. There he practised an ingenious deception upon his pursuers. Having hardly any men left in the city, for the youth of Murcia had fallen in the field, he made the women put on male attire, arm themselves with helmets and long rods like lances, and bring their hair over their chins as though they wore beards. Then he lined the ramparts with this strange garrison, and when the enemy approached in the shades of evening, they were disheartened to see the walls so well defended. Theodemir then took a flag of truce in his hand, and put a herald's tabard on his page, and they two sallied forth to capitulate, and were graciously received by the Moslem general, who did not recognize the prince. "I come," said Theodemir, "on behalf of the commander of this city to treat for terms worthy of your magnanimity and of his dignity. You perceive that the city is capable of withstanding a long siege; but he is desirous of sparing the lives of his soldiers. Promise that the inhabitants shall be at liberty to depart unmolested with their property, and the city will be delivered up to you to-morrow morning without a blow; otherwise we are prepared to fight until not a man be left." The articles of capitulation were then drawn out; and when the Moor had affixed his seal, Theodemir took the pen and wrote his signature. "Behold in me," said he, "the governor of the city!" At the dawn of day the gates were thrown open, and the Moslems looked to see a great force issuing forth, but beheld merely Theodemir and his page, in battered armour, followed by a multitude of old men, women, and children. "Where are the soldiers," asked the Moor, "that I saw lining the walls last evening?" "Soldiers have I none," answered Theodemir. "As to my garrison, behold it before you. With these women did I man my walls; and this page is my herald, guard, and retinue!" So struck was the Moorish general with the boldness and ingenuity of the trick which had been played upon him, that he made Theodemir governor of the province of Murcia, which was ever afterwards known in Arabic as "Theodemir's land." Even in these early days the Moors knew and practised the principles of true chivalry. They had already won that title to knightliness which many centuries later compelled the victorious Spaniards to address them as "Knights of Granada, Gentlemen, albeit Moors:"

> Caballeros Granadinos
> Aunque Moros hijos d'algo.

Meanwhile Tārik had pressed on to Toledo, the capital of the Goths. He was seeking for the Gothic nobles. At Cordova he had looked to meet them, but they had fled: at Toledo, which the Jews delivered into his hands, the nobles were not to be found; they had fled further, and taken refuge in the mountains of the Asturias. Traitors, like the family of Witiza and Count Julian, alone remained, and these were rewarded with posts of government. The rest of the nobility had disappeared; the country was abandoned to the Moors. Spain had become, in fact, a province of the vast empire of the Arab Khalifs,

PUERTO DEL SOL, TOLEDO.

who held their court at Damascus and swayed an empire that stretched from the mountains of India to the pillars of Hercules. What remained to be done towards the pacification of Spain was effected by Mūsa, who, when he heard of Tārik's continued career of success, sailed in all haste across the Straits, followed by his Arabs, to take his full share of the glory. He crossed in the summer of 712 with eighteen thousand men, and, after reducing Carmona, Seville, and Merida, joined Tārik at Toledo. The meeting between the conqueror and his superior officer was not friendly. Tārik went forth to receive the governor of the West with all honour, but Mūsa struck him with a whip, overwhelmed him with reprimands for exceeding his instructions, and, declaring that it was impossible to entrust the safety of the Moslems to such rash and impetuous leading, threw him into prison. When this act of jealous tyranny came to the ears of the Khalif Welīd he summoned Mūsa to Damascus, and restored Tārik to his command in Spain.

Before returning to Syria, Mūsa had stood upon the Pyrenees and seen a vision of European conquest. His recall interrupted his further advance; but others soon pushed forward. An Arab governor, as early as 719, occupied the southern part of Gaul, called Septimania, with the cities of Carcasonne and Narbonne, and from these centres he began to make raids upon Burgundy and Aquitania. Eudes, Duke of Aquitania, administered a total defeat to the Saracens under the walls of Toulouse in 721, but this only diverted their course more to the west. They sacked Beaune, exacted tribute from Sens, seized Avignon in 730, and made numerous raids upon the neighbouring districts. The new governor of Narbonne, Abd-er-Rahmān, resolved upon the conquest of all Gaul. He had already checked the operations of Eudes, who presumed, after his victory at Toulouse, to carry the war into the Saracens' country; and now he attacked the Tarraconaise, and boldly invaded Aquitaine, defeated Eudes on the banks of the Garonne, captured Bordeaux by assault, and in 732 marched on in triumph towards Tours, where he had heard of the treasures of the Abbey of St. Martin. Between Poictiers and Tours he was met by Charles, the son of Pepin the Heristal, then virtual King of France, for the feeble Merovingian sovereign, Lothair, had no voice to oppose the will of his powerful Mayor of the Palace. The Saracens went joyfully to the fight. They expected a second field of the Guadalete, and looked to see fair France their prey from Calais to Marseilles. An issue momentous for Europe was to be decided, and the conflict that ensued has rightly been numbered among the fifteen decisive battles of the world. The question to be judged by force of arms was whether Europe was to be Christian or Mohammedan—whether the future Nôtre Dame was to be a church or a mosque—perhaps even whether St. Paul's, when it came to be built, should echo the chant of the Agnus Dei or the muttered prayers of Islam. Had not the Saracens been checked at Tours there is no reason to suppose that they would have stopped at the English Channel. But, as fate decreed, the tide of Mohammedan invasion had reached its limit, and the ebb was about to set in. Charles and his Franks were no emasculate race like the Romanized Spaniards and Goths. They were at least as hardy and valorous as the Moors themselves, and their magnificent stature gave them an advantage which could not fail to tell. Six days were spent in partial engagements, and then on the seventh came a general medley. Charles cut through the ranks of the Moslems with irresistible might, dealing right and left such ponderous blows that from that day he was called Charles Martel, "Karl of the Hammer." His Frankish followers, inspired by their leader's prowess, bore down upon the Saracens with crushing force; and the whole array of the Moslems broke and fled in utter rout. The spot was long and shudderingly known in Andalusia by the name of the "Pavement of Martyrs."

The danger to Western Europe was averted. So crushing was the disaster that the Moors of Spain never again, during all the centuries that they ruled in the south, attempted to invade France. They retained, indeed, their hold of Narbonne and the districts bordering the northern slopes of the Pyrenees for some time longer (until 797), and even ventured upon foraying raids into Provence. But here their ambition ceased. The battle of Tours had once for all vindicated the independence of France, and set a bound to the Moslem conquests. Like the swelling tide of the sea, the Saracen hordes had poured over the land; and now, through the Hammerer of the Franks, a voice had spoken: "Hitherto shalt thou come and no further, and here shall thy proud waves be stayed."

On the other hand, the kings of France were so deeply impressed with the courage of their Moslem neighbours, that, though they too delighted in occasional forays, once only did they attempt the subjugation of Spain. Charlemagne, the second Alexander,

ARCH IN THE ALJAFERIA OF ZARAGOZA.

could not contemplate with composure the immunity of the Moslem power on the other side of the Pyrenees. As a good Christian he was pledged to extirpate the infidel; and, as an imperial conqueror, the existence of the independent kingdom of Andalusia was hateful to his pride. His opportunity came at last—when the accession of the first Spanish prince of the Omeyyad stock roused the hostility of some of the factions which were always prone to revolt in Spain. Charlemagne was invited to interfere and drive out the usurper. The Spanish chroniclers make Alfonso, King of the Asturias and heir of Pelagius,[1] summon the Frankish emperor to his aid; but there is more reason to believe that the invitation came from certain disappointed Moslem chiefs, who could not brook the authority of Abd-er-Rahmān the Omeyyad, and who were ready to submit even to the sworn enemy of Islam, rather than recognize the new ruler. The moment of their appeal was propitious; Charlemagne had just completed, as he thought, the subjugation of the Saxons; their chief Wittekind had been banished, and thousands of his followers were coming to Paderborn to be baptized. The conqueror's hands were thus free to turn to other schemes of victory. It was arranged that he should invade Spain, while the factious Moslem chiefs should make diversions in his favour at three different points. Fortunately for the newly-founded dynasty of Cordova, this formidable coalition came to naught. The allies in Spain miscalculated their time, and fell to blows with one another; and when Charlemagne crossed the

[1] On Pelayo or Pelagius, see below, ch. vii.

Pyrenees in 777, he found himself unsupported. He began the siege of Zaragoza, when news was brought him that Wittekind had returned and raised the Saxons, who were again in arms, and had advanced as far as Cologne. There was nothing for it but to hurry back and defend his dominions. He rapidly retraced his steps, and the main part of his army had already crossed the mountains when disaster overtook the rear in the Pass of Roncesvalles. The Basques, who nourished an eternal hatred against the Franks, had laid a skilful ambuscade among the rocky defiles of the Pyrenees, and, allowing the advanced part of the army to march through, waited till the rear-guard, encumbered with baggage, began slowly to thread its way through the pass. Then they fell upon it hip and thigh, so that scarcely a Frank escaped. The Christian chroniclers tell terrible tales of the slaughter done that day. According to them it was the Saracens, side by side with the knights of Leon, who wrought this havoc upon King Charles. We read in the old Spanish ballad how the legendary hero Bernardo del Carpio led the chivalry of Leon to the massacre of the Frankish host:

With three thousand men of Leon from the city Bernard goes,
To protect the soil Hispanian from the spear of Frankish foes;
From the city which is planted in the midst between the seas,
To preserve the name and glory of old Pelayo's victories.

Free were we born, 'tis thus they cry, though to our king we owe
The homage and the fealty behind his crest to go:
By God's behest our aid he shares, but God did ne'er command
That we should leave our children heirs of an enslavèd land.

Our breasts are not so timorous, nor are our arms so weak,
Nor are our veins so bloodless, that we our vow should break,

To sell our freedom for the fear of prince or paladin :
At least we'll sell our birthright dear—no bloodless prize they'll win.

At least King Charles, if God decrees he must be Lord of Spain,
Shall witness that the Leonese were not aroused in vain :
He shall bear witness that we died as lived our sires of old—
Nor only of Numantium's pride shall minstrels' tale be told.

The LION that hath bathed his paws in seas of Lybian gore,
Shall he not battle for the laws and liberties of yore ?
Anointed cravens may give gold to whom it likes them well,
But steadfast heart and spirit, Alfonso ne'er shall quell.

Side by side with the doughty warriors of Leon, who thus refused to join the Prince of the Asturias in his homage to Charlemagne, were (according to the romances) a host of valiant Saracens, who joined in the onset upon the retiring Franks. Pseudo-Turpin's legendary history of Charles and Orlando tells of a "fresh body of thirty thousand Saracens, who now poured furiously down upon the Christians, already faint and exhausted with fighting so long, and smote them from high to low, so that scarcely one escaped. Some were transpierced with lances, some killed with clubs, others beheaded, burnt, flayed alive, or suspended on trees." The massacre was horrible, and the memory of that day has never faded from the imagination of the peasantry of the district. When the English army pursued Napoleon's marshals through the pass of Roncesvalles, the soldiers heard the people singing the old ballad of the fatal field ; and Spanish minstrels have recorded many incidents, true or false, of the fight. One of the most famous is the ballad of Admiral Guarinos, which Don Quixote and Sancho Panza heard sung at Toboso, according to the veracious history of Cervantes :

The day of Roncesvalles was a dismal day for you,
Ye men of France, for there the lance of King Charles was broke in two:
Ye well may curse that rueful field, for many a noble peer
In fray or fight the dust did bite beneath Bernardo's spear.

There captured was Guarinos, King Charles's Admiral:
Seven Moorish kings surrounded him, and seized him for their thrall.

And the ballad goes on to tell the tale of Guarinos' captivity, and of his revenge at the tourney, when he slew his captor, and rode free for France.

Among the slain that day was Roland, the redoubtable Paladin, commander of the frontier of Brittany. He is the Sir Launcelot of the Charlemagne romance, and many are the doughty deeds recorded of him. He had fought all day in the thickest of the fray, dealing deadly blows with his good sword Durenda; but all his prowess could not save the day. So, wounded to death, and surrounded by the bodies of his friends, he stretched himself on the ground, and prepared to yield up his soul. But first he drew his faithful sword, than which he would sooner have spared the arm that wielded it, and saying, "O sword of unparalleled brightness, excellent dimensions, admirable temper, and hilt of the whitest ivory, decorated with a splendid cross of gold, topped by a berylline apple, engraved with the sacred name of God, endued with keenness and every other virtue, who now shall wield thee in battle, who shall call thee master? He that possessed thee was never conquered, never daunted by the foe; phantoms never appalled him. Aided by the Almighty, with thee did he destroy the Saracen, exalt the faith of Christ, and win consummate glory. O happy sword, keenest of the keen, never was one like thee ; he that made thee, made not thy fellow! Not one escaped with life from thy stroke." And lest Durenda should fall into the hands of a craven or an infidel, Roland smote it upon a block of stone and brake it in twain. Then he blew his horn, which was so resonant that all other horns were split by its sound ; and now he blew it with all his might, till the veins of his neck burst. And the

> blast of that dread horn,
> On Fontarabian echoes borne,

reached even to King Charles's ear as he lay encamped and ignorant of the disaster that had befallen the rear-guard eight miles away. The king would have hastened to answer the forlorn blast, that seemed to tell of a tragedy; but a traitor told him that Roland was gone a-hunting, and Charlemagne was persuaded not to answer the summons of his faithful paladin; who, after prayer and confession, gave up the ghost. Then Baldwin, another of the peers of France, came running to the king and told him of what had befallen the rear of his army, and the death of Roland and Oliver. Whereupon the king and all his army turned and marched back to Roncesvalles, where the ground was strewn with dead, and Charles himself was the first to descry the body of the hero, lying in the form of a cross, with his horn and broken sword beside him. Then did Great Charles lament over him with bitter sighs and sobs, wringing his hands and tearing his beard, and crying, "O right arm of thy Sovereign's body, honour of the Franks, sword of justice, inflexible spear, inviolable breastplate, shield of safety, noble defender of the Christians, scourge of the Saracens, a wall to the clergy, the widow's and orphan's friend, just and faithful in judgment! Renowned Count of the Franks, valiant captain of our armies, why did I leave thee here to perish? How can I behold thee dead, and not die with thee? Why hast thou left me sorrowful and alone, a poor miserable king? But thou art exalted to the kingdom of heaven, and dost enjoy the company of angels and martyrs!" Thus did Charles mourn for Roland to the last day of his life. On the spot where he died the army rested, and the body was embalmed with balsam, aloes, and myrrh. The whole army of the Franks watched by it that night, honouring the corse with hymns and songs, and lighting fires on the mountains round about. Then they took him with them, and buried him right royally. Thus ended the fatal day—

> When Roland brave and Oliver,
> And every paladin and peer,
> On Roncesvalles died.

No action of so small importance has ever been made the theme of so many heroic legends and songs. It is the Thermopylæ of the Pyrenees, with none of the glory or the significance, but all the glamour, of its prototype.

III.

THE PEOPLE OF ANDALUSIA.

THE victory of Charles Martel, in 733, had set a bound to the Saracens' invasion of Europe; they no longer thought of further conquest, but turned to the work of consolidating the kingdom they had acquired. After the brief and disastrous incursion of Charlemagne, they were left in almost undisturbed possession of their new territory for a period of three hundred years. It is true the descendants of the expelled Goths still held out in stubborn independence in the mountainous districts of the north, and from time to time recovered a portion of their ancient dominion; but these inroads, while they gave some trouble, did not materially endanger the domination of the Moors over the greater part of Spain until the eleventh century. The conquerors accepted the independence of the northern provinces as an inevitable evil, which would cost more blood to remove than the feat was worth; and leaving Galicia, Leon, Castile, and the Biscayan provinces to the Christians, they contented themselves with the better part of the land: the Christians might enjoy the dreary wastes and rocky defiles of the north, provided they did not interfere with the Moors' enjoyment of the warm and fertile provinces of the south and east. From the end of the eighth century, when the Moorish boundaries took a tolerably final shape, to the time of the advance of the Christian kingdoms in the eleventh century, the division between the Christian north and the Moslem south may be roughly placed at the great range of mountains called the Sierra de Guadarrama, which runs in a north-easterly direction from Coimbra in Portugal to Zaragoza, from whence the Ebro may be taken as a rough boundary. The Moors thus enjoyed the fertile valleys of the Tagus, the Guadiana, and the Guadalquivir — the very name of which bears witness to its Arab owners, for Guadalquivir is a corruption of the Arabic Wady-l-kebīr, or the "Great River"—besides possessing the famous cities of Andalusia, the wealth and commerce and climatic advantages of which had been celebrated from Roman times. The division was a natural one; the two parts have been distinguished geographically from time immemorial, on account of their climatic differences. The north is bleak and exposed to biting winds, subject to heavy rains and intense cold; a good pasturage country, but in most parts ill to cultivate. The south, while tormented by the hot winds that blow over from Africa, is genial, well watered, and capable of high cultivation. A great plateau divides the two, and though this fell chiefly on the Moorish side, it was to some extent debatable land and insecurely held. Its chilly heights rendered it distasteful to lovers of sunshine like the Moors, and they confided it chiefly to the care of the Berber tribes who had first come over with Tārik, and who

ALCANTARA.

were always held in poor estimation by the true Arabs who reaped the fruits of the conquest.

In the two-thirds of the peninsula thus marked off by nature for their habitation, which the Arabs always called "Andalus," and we shall call Andalusia, to distinguish it from the entire peninsula, the Moors organized that wonderful kingdom of Cordova which was the marvel of the Middle Ages, and which, when all Europe was plunged in barbaric ignorance and strife, alone held the torch of learning and civilization bright and shining before the Western world. It must not be supposed that the Moors, like the barbarian hordes who preceded them, brought desolation and tyranny in their wake. On the contrary, never was Andalusia so mildly, justly, and wisely governed as by her Arab conquerors. Where they got their talent for administration it is hard to say, for they came almost direct from their Arabian deserts, and the rapid tide of victories had left them little leisure to acquire the art of managing foreign nations. Some of their counsellors were Greeks and Spaniards, but this does not explain the problem; for these same counsellors were unable to produce similar results elsewhere, and all the administrative talent of Spain had not sufficed to make the Gothic domination tolerable to its subjects. Under the Moors, on the other hand, the people were on the whole contented—as contented as any people can be whose rulers are of a separate race and creed,—and far better pleased than they had been when their sovereigns belonged to the same religion as that which they nominally professed. Religion was, indeed, the smallest difficulty which the Moors had to contend with at the outset, though it became troublesome afterwards. The Spaniards were as much pagan as Christian; the new creed promulgated by Constantine had made little impression among the general mass of the population, who were still predominantly Roman. What they wanted was, not a creed, but the power to live their lives in peace and prosperity. This their Moorish masters gave them.

At first of course there was a brief period of confusion, some burning, pillaging, massacring; but this was soon checked by the Arab governors. When things had settled down again, the subject populations found themselves at least no worse off than before, and they shortly began to perceive that they had benefited by the change of rulers. They were permitted to retain their own laws and judges; governors of their own race administered the districts, collected the taxes, and determined such differences as arose amongst themselves. The citizen classes, instead of bearing the whole burden of the State expenditure, had only to pay a poll-tax of no very exacting amount, and they were free of all obligations; unless they held cultivable land, in which case they paid the *Kharaj* or land-tax as well. The poll-tax was graduated according to the rank of the payer, from twelve to forty-eight *dirhems* a year, or from about three to twelve pounds at our present purchasing power of money; and its collection in twelve monthly instalments made it the easier to meet. The poll-tax was an impost upon heresy; it was levied only upon Christians and Jews: the land-tax, on the other hand, which varied accord-

THE SIERRA NEVADA.

ing to the productiveness of the soil, was assessed equally on Christians, Jews, and Moslems. As a rule the old proprietors and cities preserved their property as before the conquest. The lands of the Church, indeed, and of those landowners who had fled to the mountains of the north, were confiscated, but even then their serfs were left upon them as cultivators, and were only required to pay a certain proportion, varying from a third to four-fifths, of the produce, to their new Moslem lords. Sometimes the cities, such as Merida and Orihuela, had been able to obtain exceptionally favourable terms from the conquerors, and were suffered to retain their goods and lands upon the payment of a fixed tribute. At the worst, beyond the poll-tax, the Christians were in no way subject to heavier exactions than their Moslem neighbours. They had even gained a right which had never been permitted them by the Gothic kings: they could alienate their lands.[1] In religious toleration they had nothing to regret. Instead of persecuting them, and forcing upon them a compulsory conversion, as the Goths had upon the Jews, the Arabs left them free to worship whom or what they pleased; and so valuable was the poll-tax to the treasury, that the Sultans of Cordova were much more disposed to discourage than to welcome any considerable missionary fervour that might deprive the State of so useful a source of revenue. The result was that the Christians were satisfied with the new *régime*, and openly admitted that they preferred the rule of the Moors to that of the Franks or Goths. Even their priests, who had lost most of all, were at

[1] Dozy: Hist. des Musulmans d'Espagne, livre ii. ch. ii.

first but little incensed with the change, as the old chronicle, ascribed to Isidore of Beja, written at Cordova in 754, shows. The good monk is not even scandalized at so unholy an alliance as the marriage between Roderick's widow and the son of Mūsa. But the best proof of the satisfaction of the Christians with their new rulers is the fact that there was not a single religious revolt during the eighth century.

Above all, the slaves, who had been cruelly ill-used by the Goths and Romans, had cause to congratulate themselves upon the change. Slavery is a very mild and humane institution in the hands of a good Mohammedan. The Arabian Prophet, while unable to do away with an ancient institution, which was nevertheless repugnant to the socialistic principles of Islam, did his utmost to soften the rigours of slavery. "God," said he, "hath ordained that your brothers should be your slaves: therefore him whom God hath ordained to be the slave of his brother, his brother must give him of the food which he eateth himself, and of the clothes wherewith he clotheth himself, and not order him to do anything beyond his power. . . . A man who ill-treats his slave will not enter into Paradise." There is no more commendable action in Mohammedan morals than to free slaves, and such enfranchisement is enjoined by the Prophet especially as an atonement for an undeserved blow or other injustice. In Andalusia, the slaves upon the estates that had passed from the Christians into the possession of Moslems were almost in the position of small farmers; their Mohammedan masters, whose trade was war, and who despised heartily such menial occupations as tilling the soil, left them free to cultivate the land as they pleased, and only insisted on a fair return of products. Slaves of Christians, instead of being hopelessly condemned to servitude for all their lives, were now provided with the simplest possible road to freedom : they had only to go to the nearest Mohammedan of repute, and repeat the formula of belief, "There is no god but God, and Mohammed is His Prophet," and they became immediately free. Conversion to Islam thus carried with it enfranchisement, and it is no wonder that we find the Spanish slaves hastening to profess the new faith and thus to become free men. The Catholic priests had taken small pains to graft the Christian religion into their hearts ; they had enough to do to look after their estates and the souls of the nobles without troubling themselves about the spiritual wants of the ignorant ; and the change from semi-pagan, semi-Christian, vacuity to a perhaps equally unintelligent apprehension of Islam was no very severe wrench to the servile mind. Nor were the slaves by any means the only converts to the new religion. Many of the large proprietors and men of position became Mohammedans, either to avoid the poll-tax, or to preserve their estates, or because they honestly admired the simple grandeur of this latest presentment of theism. These converts or renegades were destined to cause some trouble in the State, as will presently be seen. While admitted to the equality involved in conversion, they were not really allowed equal rights and privileges ; they were excluded from the offices of State, and regarded with suspicion by the Moslems *de la vielle roche* as interested converts, people who would sell their souls for pelf. In the end these distinctions died out, but not before they had produced serious dissensions and even insurrections.

As far as the vanquished were concerned, we have seen that the conquest of Andalusia by the Arabs was on the whole a benefit. It did away with the overgrown estates of the great nobles and churchmen, and converted them into small proprietorships; it removed the heavy burdens of the middle classes, and restricted the taxation to the test-tax per poll levied on unbelievers, and the land-tax levied equally on Moslem and Christian; and it induced a wide-spread emancipation of the slaves, and a radical improvement in the condition of the unemancipated, who now became almost independent farmers in the service of their non-agricultural Mohammedan masters.

It was otherwise with the victors. There is no greater mistake than to imagine that the Arabs, who spread with such astonishing rapidity over half the civilized world, were in any real sense a united people. So far was this from being the truth, that it demanded all Mohammed's diplomatic skill, and all his marvellous personal prestige, to keep up a semblance of unity even while he was alive. The Arabs were made up of a number of hostile tribes or clans, many of whom had been engaged in deadly blood-feuds for several generations, and all of whom were moved by a spirit of tribal jealousy which was never entirely extinguished. Had the newly-founded Mohammedan State been restrained within the borders of Arabia, there can be no doubt that it would speedily have collapsed in the rivalry of the several clans; as it was, the death of the Prophet was followed by a general rising of the tribes. Islam became a permanent and world-wide religion only when it clothed itself with armour and became a church militant. The career of conquest saved the faith. The Arabs laid aside for awhile their internecine jealousies, to join together in a grand chase for booty. There was of course a strong fanatical element in the enthusiasm of conquest. They fought partly because they were contending with the enemies of God and His Prophet, because a martyr's Benjamin's cup of happiness awaited those who fell in "the path of God," as they termed the religious war; but there is no denying that the riches of Cæsars and Chosroes, the fertile lands and prosperous cities of the neighbouring kingdoms, formed a very large element in the Moslems' zeal for the spread of the faith.

As soon as the career of conquest was exchanged for the quiet of settled possession, the various jealousies and dissensions which the tumult and profits of invasion had kept to some degree in abeyance broke forth into dangerous activity. The party spirit of the Arab tribes extended to all parts of the vast empire they had subdued, and influenced even the Khalif at Damascus; the nomination of the governors of the most distant provinces was actuated by mere factious motives. In Spain, where the "Emīr of Andalus," as he was styled, was appointed either by the Governor of Africa or by the Khalif of Damascus himself, these party differences worked havoc with the peace and order of the kingdom during the first fifty years of Moorish rule. Governors were appointed, deposed, or murdered, in deference to the mandates of some faction, who resented the government being entrusted to a man of the Medīna faction, or would not have a clansman of Kays, or objected to the nomination of a member of the Yemen party; and, throughout the history of the domination of the Moors in Spain, these baleful influences continued to work injury to the State.[1]

In Andalusia, moreover, there was another and very important party to be reckoned with, besides the various Arab factions. The conquest of the peninsula had been effected almost entirely by Tārik and his Berbers, and these Berbers (who are the *Moors* proper, though the word is conveniently employed to denote the mixture of Arabs and Berbers) formed a leading factor in the new state of things. They were not an effete nation like the Romanized Spaniards; but a people full of life and martial energy. In their mountain fastnesses, and ranging the plains from Egypt to the Atlantic, in their numerous and widely distinguished clans, the Berbers had offered to the Arabs a much more formidable resistance than the trained soldiers of Persia or Rome. In many ways they resembled their invaders: they were clansmen like the Arabs; their political ideas were democratic like theirs, with the same reverence for noble families, which took away the dangerous qualities of pure democracy among an ignorant people. Their very manner of warfare was almost Arab. For seventy years the two races of nomads fought together, and

[1] Dozy Hist. des Mus. d'Espagne, livre i.

when at last the Arabs obtained the upper hand, it was rather by the acquiescence of their foes than by any distinct submission. The Berbers permitted the Arab governor to hold his court near the coast, but insisted on preserving their own tribal government among themselves, and demanded to be treated as brothers, not as servants, by their antagonists. This fraternal system worked fairly well for a time. The Berbers, always a marvellously credulous people, were quick to accept any new faith, and embraced Islam with a fervour far exceeding anything the more sceptical mind of the Arab could evoke. Very soon Barbary became the hotbed of religious nonconformity; the arid doctrines of Islam were supplemented by those more mystical and emotional elements which imaginative minds soon engraft upon any creed soever; and the Mohammedan dissenter, expelled from the more rigid regions of orthodoxy, found a singularly productive soil for his doctrines in the simple minds of the Berbers. The same susceptibility to religious emotion, which had produced so general a conversion that the conquest of Spain was effected by a Berber general and twelve thousand Berber troops, soon led to further movements. The *Marabout*—saint, missionary, or priest—came to exercise a more potent influence over this credulous people than tribal chief or Arab governor could ever acquire. It needed but a few mock miracles to bring a host of gaping devotees about the shrine of the *marabout*, and so clearly had an Arab general realized this condition of popularity that, when he perceived the influence which a priestess exercised over the people by her jugglery, the subtle Moslem set to work in the same manner, and soon became an adept at legerdemain or whatever corresponded to spirit-rapping in those days, with the very best results. But a people so easily influenced by such means, a priest-ridden nation, is always liable to sudden and violent revolutions, which its priests can stimulate by a single word. The *marabouts* among the Berbers were responsible for most of the later changes that took place in North Africa: they set up the Fatimites, sent the Almoravides victorious through Barbary and Spain, and then put them down by the Almohades. They began very early to work against the Arab governors, and when one of these had indulged his passion for luxury at the expense of a cruel oppression of his subjects, the priests set the Berbers in revolt, and in a moment the whole of the western half of the Mediterranean coast was up in arms, and the Arabs were terribly defeated. Thirty thousand fresh troops were sent from Syria to recover the provinces, but these, joined to the Arabs that still remained in Africa, were repulsed with great slaughter, and the remnant were cooped up in Ceuta, where they daily awaited famine and massacre.

The Berbers in Andalusia, always in intimate touch with their kinsmen over the water, were quick to feel the influence of such a revolution as was then (741) going forward in Africa. They had cause to grudge the Arabs their lion's share of the spoils of Spain, which had been the trophies of the Berbers' bow and spear. While the Arabs, who had only arrived in time to reap the advantages of the conquest, had appropriated all the most smiling provinces of the peninsula, the Berbers found themselves relegated to the most unlovely parts, to the dusty plains of Estremadura, or to the icy mountains of Leon, where they had to contend with a climate which severely tried natures brought up in African heats, and where, too, they had the doubtful privilege of forming a buffer between their Arab allies and the Christians of the North. Already there had been signs of disaffection. One of Tārik's Berber generals, Monousa, who had married a daughter of Eudes, Duke of Aquitaine, raised the standard of revolt when he heard of the oppression of his countrymen in Africa; and now, when the Berber cause was triumphant across the Straits, a general rising took place among the northern provinces; the Berbers of the borders, of Galicia, of Merida, Coria, and all the region round about, took up arms, and began to march south upon Toledo, Cordova, and Algeciras, whence they intended to take ship and go to join their compatriots in Barbary.

The situation was full of peril, and the Arab Emīr of Andalusia, Abd-el-Melik, who had sternly refused to lend any assistance to the Syrian Arabs shut up in Ceuta, now found himself in this dilemma, that either he must submit to his own rebellious Berbers, or he must invite the co-operation of the very Syrians whom he had persistently refused to succour, and who, when they arrived, might possibly turn out to be a worse plague than that they came to remove. In grave apprehension, he sent ships and brought over the Syrians, after first making them promise to go back when their work was done. Thus reinforced, the Arabs of Andalusia put the Berbers to utter rout, hunted them like wild beasts through the country to their mountain fastnesses, and gratified their vengeance to the full. And then the event which Abd-el-Melik had endeavoured to guard against came to pass. The Syrian auxiliaries refused to exchange the rich lands of Andalusia for the deserts of Africa and the spears of triumphant Berbers; they defied and murdered Abd-el-Melik, and set up their own chief in his stead. The result was a long and obstinate struggle between the old Arab party and the new-comers, accompanied by much bloodshed and devastation. The struggle was only decided when the Khalif of Damascus sent over a new and able governor, who divided the hostile factions by giving them settlements in cities far apart from each other, and banished the more turbulent of their leaders. Thus the Egyptian contingent of the Syrian army was settled in Murcia, which they re-christened "Misr" or Egypt; the men of Palestine at Sidonia and Algeciras; the people of the Jordan at Regio (Malaga), those of Damascus in Elvira (Granada), and the battalion of Kinnesrin at Jaen.[1] From this time one of the causes of faction in Andalusia was removed, but party spirit still ran high, and government was often changed to anarchy, until a ruler armed with peculiar prestige, carrying in his person the authority and blood of the Khalifs of Damascus, came to take into his hands the sceptre of

[1] Makkary: History of the Mohammedan Dynasties in Spain (Gayangos), vol. ii. p. 46. Dozy: Hist. des Mus. d'Espagne, livre i. ch. xii.

the disturbed country and to unite for awhile all factions under the standard of the Sultan of Cordova. This young man was the new ruler whom Charlemagne had so unsuccessfully come to expel, and his name was Abd-er-Rahmān the Omeyyad.

IV.

A YOUNG PRETENDER.

FOR six hundred years the greater part of the Mohammedan Empire was nominally under the authority of a central ruler called a Khalif, a title which signifies a "successor" or "substitute." At first this authority was real and powerful: the Khalif appointed the governors of all the provinces, from Spain to the borders of the Hindu Kush, and removed any of them at his pleasure. But the empire was too large to hold together round a central pivot for any length of time, and gradually various local governors made themselves virtually independent, although they generally professed the utmost devotion to the Khalif and paid him every honour except obedience. By degrees even this show of respect was thrown off, and dynasties arose which espoused heretical tenets, repudiated the spiritual supremacy of the Khalif, and denounced him and all his line as usurpers. Finally the time came when the Khalifs were as weak in temporal authority as the Pope of Rome, and were even kept prisoners in their palace by the mercenary bodyguard they had hired to protect them against their rebellious nobles. This took place about three hundred years after the foundation of the Khalifate;

and for the second half of their existence the Khalifs were little more than ciphers to be played with by the great princes of the empire and to contribute a little pomp to their coronations. Finally the Khalifate was abolished in Asia by the Mongol invasion in the thirteenth century, and though the title is still claimed by the Sultan of Turkey, there is no Khalif now in the old comprehensive sense of the word.[1]

The earliest province to shake off the authority of the Khalif was Andalusia. To understand how this happened, we must remember that the Khalifs did not succeed one another in one unbroken line of family inheritance. After the first four (or "orthodox") Khalifs, Abu-Bekr, Omar, Othmān, and Aly, who were elected more or less by popular vote, the Syrian party set up Moāwia as Khalif at Damascus, and from him sprang the family of the Omeyyad Khalifs, so called from their ancestor Omeyya. There were fourteen Omeyyad Khalifs, who reigned from 661 to 750, when they were deposed by Es-Seffāh, "the Butcher," who was the first of the second dynasty of Khalifs, called Abbāside, after their ancestor Abbās, an uncle of the Prophet Mohammed. The Abbāside Khalifs transferred the seat of government from Damascus to Baghdad, and held the Khalifate until its destruction by the Mongols in 1258. Among the members of the deposed family of the Omeyyads was Abd-er-Rahmān, a name which means "Servant of the Merciful God." Most of his relations were

[1] For an account of the power of the body-guard and the fall of the Khalifate, the reader is referred to The Story of the Saracens, by Arthur Gilman.

exterminated by the ruthless Abbāside; they were hunted down in all parts of the world and slain without mercy. Abd-er-Rahmān fled like the rest, but with better fortune, for he reached the banks of the Euphrates in safety. One day, as he sat in his tent watching his little boy playing outside, the child ran to him in affright, and, going out to discover the cause, Abd-er-Rahmān saw the village in confusion, and the black standards of the Abbāsides on the horizon. Hastily seizing up his child, the young prince rushed out of the village, and reached the river. Here the enemy almost came up with them, and called out that they need have no fear, for no injury would be done to them. A young brother, who had accompanied him, and who was exhausted with swimming, turned back, and his head was immediately severed from his body; but Abd-er-Rahmān held on till he reached the other side, bearing his child, and followed by his servant Bedr. Once more on firm earth, they journeyed night and day till they came to Africa, where the rest of his family joined them, and the sole survivor of the Omeyyad princes had leisure to think of his future.

He was but twenty years of age, and full of hope and ambition. His mental powers were considerable, and to these he added the advantages of a noble stature and great physical energy and courage. The Arab historians, however, add the unfavourable details that he was blind of one eye and devoid of the sense of smell. In his childhood wise men had predicted great things of his future, and in spite of the ruin of his family he was not yet daunted. His first thoughts turned to Africa; for he clearly perceived that the success of the Abbāsides had left him no chance in the East. But after five years of wandering about the Barbary coast he realized that the Arab governor was not easily to be overturned, and that the already revolted Berbers in the West would not willingly surrender their newly-won independence for the empty glory of being ruled by an Omeyyad. His glance, therefore, was now directed towards Andalusia, where the various factions, in their perpetual strife, offered an opening to any clever pretender, and much more to one who could bring such hereditary claims as Abd-er-Rahmān. He therefore sent his servant Bedr to the chiefs of the Syrian party in Spain, among whom many were freedmen of the Omeyyads and were thus bound by the Arab code of honour to succour any relation of their former patrons. Bedr found these chiefs willing to receive the young prince, and, after some negotiation with the hostile factions, the support of the men from the Yemen was also promised. Upon this Bedr returned to Africa.

Abd-er-Rahmān was saying his prayers on the seashore when he saw the vessel approaching which brought him the good news; and, prone as all Easterns are to draw omens from insignificant circumstances, the name of the first envoy from Andalusia who was presented to him, Abu-Ghālib Temmām (which means Father of Conquest Attainment) suggested a happy fate: "We shall *attain* our object," cried the prince, "and *conquer* the land!" Without delay he stepped on board, and they sailed for Spain in September, 755. The coming of the survivor of the Omeyyads to Andalusia was like a page of romance, like the arrival of the Young Pretender in Scotland in 1745. The news spread like a conflagration through the land; the old adherents of the royal family hurried to pay him homage; the descendants of the Omeyyad freedmen put themselves under his orders. Even the Yemen clans, though they could not be expected to feel any peculiar sentiment for the young prince, were sufficiently infected by the zeal of his adherents to keep to their promise and band together for his support. The Governor of Andalusia found himself deserted by most of his troops and forced to wait for a new army; and meanwhile the winter rains made a campaign impossible, and left Abd-er-Rahmān leisure to recruit and organize his forces.

In the spring of the following year the struggle began in earnest. Abd-er-Rahmān was received with enthusiasm at Archidona and Seville, and thence prepared to march on Cordova. Yūsuf, the governor, advanced to resist him, but the Guadelquivir was swollen with rains, and the two armies, on opposite banks, raced with each other who should first arrive at Cordova. At length Abd-er-Rahmān, by means of a deceitful stratagem, unworthy of a prince of romance, induced Yūsuf to let him cross the now falling river under pretext of peace; and once on the other side, he fell upon the unsuspecting enemy. Victory declared itself for the prince, and he entered Cordova in triumph. He had the grace to exert himself to arrest the plundering passions of his troops, and to place the harīm or women-folk of the ex-governor in safety. Before the year was out he was master of all the Mohammedan part of Spain, and the dynasty of the Omeyyads of Cordova, destined to endure for nearly three centuries, was established.[1]

The King of Cordova, however, was not firmly seated without many a struggle. Abd-er-Rahmān had indeed been placed on the throne, but the feat had been accomplished by a small faction out of the numerous parties that divided the land. The new Sultan was, however, better able than most princes to hold his own amidst the striving elements of his kingdom. Prompt and decisive in action, troubled by few scruples, by turns terribly severe and perfidiously diplomatic, his policy was always equal to an emergency ; and there were not a few occasions on which it was put to the test. He had not been long in Andalusia when Ibn-Mughīth sailed from Africa to set up the black standards of the Abbāsides in Spain. He landed in the province of Beja, and soon found supporters among the disaffected, always ready to join in some new thing. Abd-er-Rahmān was besieged for two months in Carmona. The situation was perilous in the extreme, for every day gave the enemy more opportunity of increasing their forces. Abd-er-Rahmān, ever full of resource, hearing that the enemy had somewhat relaxed their precautions, gathered together seven hundred of his bravest followers, kindled a great fire, and, saying that it was now a question of death or victory, flung his scabbard

[1] Dozy : Hist. des Mus. d'Espagne, livre i. ch. xiii.–xvi.

into the flames. The seven hundred followed his example, in token of their resolution never to sheathe their swords again till they were free, and, sallying out after their leader, fell upon the besiegers tooth and nail. The Abbāside invasion was utterly annihilated. Abd-er-Rahmān, with the ferocity that occasionally disfigured him, put their leaders' heads in a bag, with descriptive labels attached to their ears, and confided the precious parcel to a pilgrim bound for Mekka, by whom it was put into the hands of the Abbāside Khalif Mansūr himself. When the Khalif had seen the contents of the bag, he was very wroth; but he could not help exclaiming, "Thank God there is a sea between that man and me!" While cordially detesting the successful Sultan of Cordova, his Abbāside foe was forced to render homage to his skill and courage. He called Abd-er-Rahmān "the hawk of the Koreysh," the falcon of the Prophet's own tribe. "Wonderful," he would exclaim, "is the daring, wisdom, and prudence, he has shown! To enter the paths of destruction, throw himself into a distant land, hard to approach, and well defended; there to profit by the jealousies of the rival parties, to make them turn their arms against one another instead of against himself; to win the homage and obedience of his subjects; and, having overcome every difficulty, to rule supreme lord of all! Of a truth, no man before him has done this!"

The defeat of the Abbāside invasion was followed by other successes on the part of the new Sultan. He induced the people of Toledo, who had long held out against him, to consent to a peace and deliver up their chiefs; and the leaders were grossly humiliated and then crucified. The chief of the Yemenite faction proving dangerous, Abd-er-Rahmān gave him a safe-conduct, and thus enticed him into his palace, where he tried to stab him with his own hand, but finding the Arab too vigorous, called in the guard and had him assassinated. Almost immediately, a great revolt of the Berbers of the northern borders occurred. Ten years were occupied in reducing them to obedience, and meanwhile the Yemenites, burning with vengeance for the murder of their chief, took advantage of the Sultan's absence in the north to rise. They had not yet realized the energy or the astuteness of the man. He had already set the revolted Berbers by the ears by playing upon their petty jealousies; and he now exerted his diplomacy to breed discord among the Yemenites. He tampered with the Berbers who formed a large part of their army, so that they deserted in the midst of the fray, and Abd-er-Rahmān's soldiers fell upon the flying multitude, until thirty thousand bodies lay on the field: their huge grave long remained a sight to be seen by the curious. Then followed that formidable coalition between three disaffected Arab chiefs and Charlemagne, which was so near destroying the fabric that Abd-er-Rahmān had painfully built up, but collapsed before Zaragoza and at Roncesvalles without a single blow from the very person they had assembled to destroy.

Henceforward the Sultan was allowed to enjoy in comparative peace the fruits of his victories. He had subdued all the hostile elements in Spain to his iron will; he had cast down the proud Arab chiefs who had dared to measure swords with him; he had massacred, or assassinated, the leaders of rebellion, and had proved himself master of the position. But tyranny, cruel and perfidious as his, brings its own punishment. The tyrant may force the submission, but he cannot compel the devotion of his people, and the empire that is won by the sword must be sustained by the same weapon. Honest men refused to enter into the service of a lord who could betray and slay as did this Sultan; his old supporters, those who had first welcomed him to Spain, now turned coldly away when they saw the tyrant in his naked cruelty; his own relations, who had flocked over to his Court, as an asylum from the Abbāsides, found his despotism so intolerable that they plotted again and again to depose him, with the inevitable result of losing their heads. Abd-er-Rahmān was left in mournful solitude. His old friends had deserted him; his enemies, though helpless, cursed him none the less; his very kinsmen and servants turned against him. It was partly that the long war with faction had spoilt a fine nature; partly that the character was relentless. No longer could he mingle as before in the crowds that thronged the streets of Cordova; suspicious of every one, wrapped in gloomy thoughts and distracted by bloody memories, he rode through the streets surrounded by a strong guard of foreigners. Forty thousand Africans, whose devotion to their paymaster was equalled by their hatred of the whole population whom they repressed, formed the Sultan's protection against the people whom he ground under his heel. In his desolation he wrote a poem on a palm which he transplanted from the land of his ancestors—for, like most Andalusian Arabs, he was something of a poet—in which he compassionated the tree for its exile: "Like me, thou art separated from relations and friends; thou didst grow in a different soil, and now thou art far from the land of thy birth." He had accomplished the object which he had set before himself in the days of his young ambition, when he came a stranger and alone to subdue a kingdom: he had brought the Arabs and Berbers into subjection, and restored order and peace in the land; but he had done it all at the expense of his subjects' hearts. The handsome youth who had come like "the young chevalier" to win the homage and devotion of the Spanish Arabs, after thirty-two years went down to his grave a detested tyrant, upheld in his blood-stained throne only by the swords of mercenaries whose loyalty was purchased by gold. He had inaugurated the sway of the sword in Spain, and his successors would have to maintain the principle. As the great historian of the Moors has observed, it is not easy to see by what other means the turbulent factions of Arabs and Berbers were to be kept in order, or how anarchy was to be averted without severe measures of repression: neither of these races was accustomed to monarchy. Nevertheless a tyranny so sustained formed a melancholy spectacle, despite all the glories and triumphs that illumined it.

An ancient Arab historian, Ibn-Hayyān, gives the following portrait of the first Sultan of Cordova:

"Abd-er-Rahmān was kind-hearted and well disposed to mercy. He was eloquent in his speech, and endowed with a quick perception. He was very slow in his determinations, but constant and persevering in carrying them into effect. He was active and stirring; he would never lie in repose, or abandon himself to indulgence. He never entrusted the affairs of government to any one, but administered them himself; yet he never failed to consult in cases of difficulty the men of wisdom and experience. He was a brave and intrepid warrior, always the first in the battle-field; terrible in his anger, and intolerant of opposition: his countenance inspired awe in those who approached him, friends and foes alike. He was wont to follow biers and pray over the dead, and in the mosque on Fridays he would often enter the pulpit and address the people. He visited the sick, and mixed with the people in their rejoicings." This is doubtless the young Abd-er-Rahmān, before opposition and conspiracy had made him suspicious and cruel. Power has often a terrible manner of punishing its possessors.

The usual question that is asked, when a despot dies, is, Who will succeed him? And the common answer is, Revolution and anarchy. A throne that is set upon steel edges does not readily pass from father to son. Yet the dynasty of Abd-er-Rahmān did not collapse with the death of its despotic founder. It was to be expected that the many hostile forces which he had with difficulty restrained, when released by his death, would have sprung into redoubled activity. Such, however, was not the case.

THE BRIDGE OF CORDOVA.

Partly because he had too thoroughly terrified the people for them easily to recover their courage, and partly because in his successor they recognized the very antithesis of his father—a prince to be loved and honoured—the people remained quiet for some years. Hishām, who in 788 succeeded his father, at the age of thirty, was a model of all the virtues; and, as if to make sure that he should practise them with assiduity during his brief reign, an astrologer predicted that he had but eight years to live. The Sultan naturally devoted this short space to preparing for the next world. In his youth his palace had been filled with men of science, poets, and sages; and the boy was father of the man. His acts of piety were numberless, and in him the indigent and the persecuted had a sure refuge. He would send trusty emissaries into all parts of his dominions to seek out wrong-doing and repress it, and to further the cause of righteousness. He had the streets patrolled at night to prevent riotous and vicious conduct; and the fines they levied on the evildoers were distributed among those good souls whom rain and cold could not deter from attending the mosques at night-time. The Sultan himself visited the sick, and would often go forth on stormy nights to carry food to some pious invalid and to watch beside his bedside. With all this he was no poltroon. He would lead his armies against the Christians of the North, like the thoroughbred Arab he was; and, though the people affectionately dubbed him "The Amiable" and "The Just," he could show sufficient firmness when his reign was menaced by the conspiracies of his uncles. He increased the number of his mamlūks, or body-guard, and a thousand of them were always on duty day and night on both sides of the river to protect his palace. He was a huntsman; yet so scrupulous was he that when he rebuilt the bridge of Cordova, which still stands to this day, hearing that his subjects murmured that he only built this great work to make his hunting parties more convenient, he vowed he would never cross it again; and he never did. Before the eight years had quite expired, this exemplary prince was gathered to his well-earned paradise; and then it became apparent that his very goodness had but served to stir up a new factor of rebellion in the State.

This new danger was the power of the Mohammedan priests. The term is hardly an accurate one, for in Islam there is no priesthood in the strict sense of Catholic Christianity. The men who recite the prayers and preach the weekly sermons in the mosques are laymen taken from their shops or other occupations, and appointed for the time to lead the congregations. There is no distinction between laic and cleric in Islam. Nevertheless, there is something which tallies more or less with what we mean by a priesthood. There is always in Mohammedan countries a body of men whose lives are specially devoted to religion; they may be dervishes with peculiar rites, or they may be merely theological students, pupils of some renowned teacher, whose doctrine fills them with unwonted zeal and enthusiasm; they may be reciters of the Koran, or schoolmasters. Such a body is found throughout the Moslem world, and it has to be reckoned with in every

Mohammedan country. The students of the Azhar mosque at Cairo, the Softas of Constantinople, the Mullas of many an Eastern city, have shown what the force of fanaticism can avail in times of excitement. In Andalusia this power was now about to be displayed. The first rebellion after Abd-er-Rahmān's death came from the least expected quarter; not from the Christians, nor from any special political party of Arabs or of Berbers, but from the devout sons of Islam, the theological students of Cordova.

These students were largely composed of renegades, or the sons of renegades. It has already been seen that the Spaniards cheerfully adopted Islam, and, like most converts, became more Moslem than the Moslems themselves. Abd-er-Rahmān was far too wise, and also far too worldly, to permit the theologians — especially those of Spanish blood — any preponderating influence in his kingdom; but the pious Hishām neither saw the danger, nor, had he perceived it, would have regarded it as a danger at all. He loved to place his confidence in holy men, whose conduct was dictated by the strict observance of their religion, and in whom he failed to detect the germs of common worldly ambition and love of power. It happened, too, that at this time the theologians were headed by a singularly gifted and active mind, a favourite pupil of one of the lights of the Holy City Medina, where the Arabian Prophet was buried, and a man whose soul was devoured by that mixture of religious fervour and political ambition which has so often made havoc of nations. This doctor, Yahya, profited by the devotion and piety of Hishām to raise the theologians of Cordova to a height of influence and power that might have made his shrewd father, Abd-er-Rahmān, turn in his grave. So long, indeed, as they had their own way, all went well. But in 796, when the good Hishām departed in the odour of sanctity, a complete change came over the Court. The new Sultan, Hakam, was not indifferent to religion or in any way a reprobate; but he was gay and sociable, and enjoyed life as it came to him, without the slightest leaning towards asceticism. Such a character was wholly objectionable to the bigoted doctors of theology. They spoke of the Sultan with pious horror, publicly prayed for his conversion, and even reviled and insulted him to his face. Finding him incurable in his levity, they plotted to set up another member of his family on the throne. The conspiracy failed, and many of the leading nobles, who had joined in the plot, together with a number of fanatical doctors, were crucified. Undeterred by this, in 806 the people, stirred up by the bigots, rose again, only to be as summarily repressed as before. Even the terrible fate of the nobles of Toledo,—who had rebelled, as was their wont, and were at this time treacherously inveigled into the hands of the Crown Prince and massacred to a man,—did not deter the Cordovans from another revolt.

For seven years, indeed, the memory of the "Day of the Foss," as the massacre at Toledo was called, kept the fanatics of Cordova within bounds; but as the recollection of that fearful hole into which the murdered bodies of all the nobility of Toledo had been cast, grew fainter, there were symptoms of a fresh insurrection at the capital. Popular feeling ran very high, not only against the Sultan, because he would not wear sackcloth and ashes or pretend to be an ascetic, but still more against his large body-guard of "Mutes," so called because, being negroes and the like, they could not speak Arabic. The Mutes dared not venture in the streets of Cordova except in numbers; a single soldier was sure to be mobbed, and might be murdered. One day a wanton blow struck by a member of the guard roused the whole people. They rushed with one accord to the palace, led by the thousands of theological students who inhabited the southern suburb of the city, and seemed bent on carrying it by assault in spite of its fortifications and garrison. The Sultan Hakam looked forth over the sea of faces, and watched with consternation the devoted mob repulsing the charge of his tried cavalry; but even in this hour of desperate peril he did not lose the *sang-froid* which is the birthright of great men. Retiring to his hall, he told his page Hyacinth to bring him a bottle of civet, with which he proceeded calmly to perfume his hair and beard. The page could not repress his astonishment at such an occupation, when the cruel mob was even then battering at the gates; but Hakam, who was fully aware of his danger, replied: "Silence, rascal! How do you suppose the rebels would be able to find out my head among the rest, if it were not distinguished by its sweet odour?" He then summoned his officers, and took his measures for the defence. These were simple enough; but they proved effectual. He despatched his cousin with a force of cavalry, by a roundabout way, to the southern suburb, which he set in flames, and when the people turned back in terror from the besieged palace to rescue their wives and children from their burning homes, Hakam and the rest of the garrison fell on them in the rear. Attacked on both hands, the unfortunate rebels were cut to pieces; the grim Mutes rode through them, slashing them down by the hundred, and disregarding, if they understood, their prayers for mercy. Hakam's manœuvre saved the palace and the dynasty; and the insurrection was converted into a wholesale massacre.[1]

Yet in the moment of his triumph the Sultan stayed his hand; he did not press his victory to the last limits, but was content with ordering the destruction of the rebellious suburb and the exile of its inhabitants, who were forced to fly, some to Alexandria, to the number of fifteen thousand, besides women and children, whence they eventually crossed to Crete; others, eight thousand in all, to Fez, in Africa. The majority of the exiles were descendants of the old Spanish population, who had embraced Islam, but were glad of a pretext to assert their racial antipathy for the Arab rule. The chief offenders, the *fakis*, or theological students, however, were left unpunished, partly, no doubt, because many of them were Arabs, and partly in deference to their profession of orthodoxy. To one of their leaders, who was dragged before Hakam, and who told the Sultan, in the heat of his fanatical rage, that in

[1] Dozy: Hist. des Mus. d'Espagne, livre ii. ch. iii., iv.

hating his king he was obeying the voice of God, Hakam made the memorable reply: "He who commanded thee, as thou dost pretend, to hate me, commands me to pardon thee. Go and live, in God's protection!"

V.

THE CHRISTIAN MARTYRS.

THE Sultan Hakam died in 822, after a reign of twenty-six years. He left a comparatively tranquil inheritance to his son Abd-er-Rahmān II.; the renegades of Cordova had been subdued and exiled, the bigots had been given a lesson that they were not likely to forget, and there only remained the chronic disturbances on the Christian borders to be occasionally repressed. Abd-er-Rahmān II. inherited his father's talent for enjoyment, but not that strength of character by which self-indulgence was preserved from degenerating into weakness. The new Sultan converted Cordova into a second Baghdad, and imitated the prodigalities of the great Harūn-er-Rashīd, who had recently left the scene of his fantastic amusements for, let us hope, a better world. Abd-er-Rahmān built palaces, laid out gardens, and beautified his capital with mosques, mansions, and bridges. Like all cultivated Moslem sovereigns, he was a lover of poetry, and claimed to be no mean poet himself, though his verses were sometimes written by other pens whom he paid to compose for him. His tastes were refined, and his nature was gentle and easily led. Four people ruled him throughout his career: one

MOORISH IVORY CASKET OF THE 11TH CENTURY IN THE CATHEDRAL OF PAMPLONA.

was a singer, the second a theologian, the third a woman, and the fourth a black slave. The most influential of these was the theologian Yahya, the same who had before stirred up the students against Hakam, and who now acquired an absolute ascendency over the mind of the new Sultan. The Queen Tarūb and the slave Nasr, however, exercised no light authority in political matters; but the singer Ziryāb confined his interest to matters of taste and culture, and refused to meddle in the vulgar strife of politics. He was a Persian, and had been a pupil of the famous musician of Baghdad, Isaac the Mosilite, until one day he had the misfortune to excel his master in a performance before the Khalif Harūn, and had immediately afterwards been offered by the jealous Mosilite the choice of death or banishment. He accepted the latter; and, arriving in Spain, was received with effusion by the cultivated Sultan, who assigned him a handsome pension, supplies of food, houses, and other privileges and allowances, so that the fortunate singer counted an immense income. So delighted was the Sultan with Ziryāb's talents that he would seat him beside him, and share his meals with him, and would listen for hours to his songs and to the wonderful tales he could tell of bygone times, and the wise sayings he could relate from his boundless stores of reading. He knew more than a thousand songs by heart, each with its separate tune, which he said the spirits of the air taught him; he added a fifth string to the lute, and his style of playing was quite unlike any one else's, so that people who had heard him would listen to none other afterwards. He had a curious way with his musical pupils. He used to make the would-be singer sit down and try to sing his loudest. If the voice was weak, he told him to tie a band round his waist to increase the volume of sound; if he stammered or had any defect in his speech, Ziryāb made him keep a piece of wood in his mouth till his jaws were properly stretched. After this, if the novice could shout *Ah* at the top of his voice, and keep the sound sustained, he took him as a pupil and trained him carefully; if not, he dismissed him.[1] Never was any one so polished, so witty, so entertaining as Ziryāb; he soon became the most popular man in Andalusia, and held the position of arbiter of fashion, like Petronius or Beau Brummell. He made the people change their manner of wearing their hair. He introduced asparagus and force-meat balls to Andalusia, and a dish was long afterwards known as "Ziryāb's fricassee." He set the example of drinking out of glass vessels instead of metal, of sleeping on leather beds, dining off leather mats, and a host of other refinements; while he insisted on a careful gradation of clothes, diminishing by slow degrees from the thick of winter to the thin of summer, instead of the abrupt change which the people had hitherto made. Whatever he prescribed, the fashionable world followed; there was nothing that this delightful epicure could not persuade them to think both necessary and charming.

But while the Court was preoccupied with the tasting of new dishes, or the cut of its hair, there

[1] Makkary: ii. 121. Dozy: livre ii. ch. v.

were earnest people among the subjects of the Sultan, in Cordova itself, who were absorbed by much deeper thoughts. It was not the external enemy that thus endangered the peace of the Moorish kingdom. Many a time, indeed, did Abd-er-Rahmān II., who was not wanting in personal courage and love of military glory, lead his armies with success against the Christians of the north, who, aided by Louis the Debonnaire, were continually making some expedition or foray over the frontiers. These petty campaigns were not yet serious enough to shake the stability of the Moslem rule. The trouble in these early days always came from within. In the present instance it arose from the too exalted spirit of a small number of Christians at Cordova. Most of the Christians, indeed, were by no means anxious to emphasize their creed; they found themselves well treated, free to worship as they pleased, with no hindrance from their rulers; and also free to trade and get rich, as well as their Moslem neighbours. What more could be desired, unless the recovery of their ancient kingdom? And as that was impossible just then, they were content to let well alone, and make the best of their mild and tolerant governors.

This temper was very general in Andalusia, but there were here and there ambitious or enthusiastic spirits that chafed against such compliance with the rule of the "infidel." They remembered the former power and prosperity of their church, and the priests especially could no longer restrain their hatred of the Moslems who had taken away from them their authority and substituted a false creed for the religion of Christ. The very tolerance of the Moors only exasperated such fervent souls; they preferred to be persecuted, like the saints of old; they longed to be martyrs, and they were indignant with the Moslems, because they would not "persecute them for righteousness' sake" and ensure them the kingdom of heaven. Especially hateful to these earnest people was the open gaiety and sensuous refinement of the Moors; their enjoyment of life and all its pleasure, their music and singing, their very learning and science, were abhorrent to these ascetics. Life, to the true believer, meant only scourges and fasts, penances and confessions, purification through suffering, the mortifying of the flesh and sanctifying of the spirit. What happened was, in truth, nothing but the manifestation of the ascetic or monastic form of Christianity among the subject populations. A sudden and violent enthusiasm took the place of the indifference that had hitherto been the prevailing characteristic of Spanish Christianity, and a race for martyrdom began.

It was a grievous pity to see good people throwing away their lives, and the lives of others, for a dream. The suicides of Andalusia were really no whit more reasonable or truly religious than the sufferings of the priests of Baal who cut themselves with knives, or of the Indian ascetics who let their nails grow through the palms of their hands. The fact that the Spanish "martyrs" were mad in a better cause does not make them less insane. Christianity does not teach its disciples to fling away their lives wantonly, out of mere joy in being tortured and killed. It was not as if the Christians were persecuted or hindered in the exercise of their faith; it was not as if the Moors were ignorant of Christianity and needed to be preached to. They knew more of the Scriptures than many of the Christians themselves, and they never spoke the name of Jesus Christ without adding, "May God bless him." Mohammedanism recognizes the inspired nature of Christ, and inculcates profound reverence towards him. The Moslems were not ignorant of Christianity, but they preferred their own creed; and while they let the Christians hold to theirs, there was no excuse for the latter posing in the heroic character of persecuted believers. Indeed there was no rational way of getting martyred; since Christians were allowed free exercise of their religious rites, might preach and teach without let or hindrance, they could not find a legal ground for being persecuted unless they left the paths of the Gospel and set aside the great lesson of Christ, "Love your enemies, do good to them that hate you, and pray for them that despitefully use you and persecute you." They were not despitefully used or persecuted; the mass of the Christians were entirely unmolested, and though the priests were sometimes subjected to some public ridicule by the street boys and common people, the better class of Moslems never joined in this; yet so far were the poor Christians from attempting to love these mild adversaries that they went out of their way to curse them and blaspheme their religion, with the simple intention of being martyred for their pains. Now it is a well-known law in Moslem countries that he who blasphemes the Prophet Mohammed or his religion must die. It is a stern and barbarous law, but the world has seen as bad principles carried into effect over the faggots of Smithfield and Oxford in later ages than that of which we are writing. Wilfully to stir up religious strife and injuriously to abuse another faith are no deeds for Christians ; voluntarily to transgress a law which carries with it capital punishment is not martyrdom, but suicide ; and the pity we cannot help feeling for the "martyrs" of Cordova is the same that one entertains for many less exalted forms of hysterical disorder. The victims were, indeed, martyrs to disease, and their fate is as pitiable as though they had really been martyrs for the faith.

The leading spirit of these suicides was Eulogius,[1] a priest who belonged to an old family of Cordova, always noted for its Christian zeal. Eulogius had spent years in prayer and fasting, in bitter penance and self-mortification, and had reduced himself to the ecstatic condition which leads to acts of misguided but heroic devotion. There was nothing worldly left in him, no thought for himself or personal ambition; to cover the false faith of the Moors with contumely, and to awaken a spirit of exalted devotion among his co-religionists, such were his aims. In these he had throughout the cordial support of a wealthy young man of Cordova. Alvaro by name, and of a small but fervid body of priests, monks, and women, with a few laymen. Among those who found a close affinity to the devoted young priest, was a beautiful girl named Flora. She was the child of a mixed marriage, and her Christian mother had brought her up secretly in

[1] Dozy: Hist. des Mus. d'Espagne, ch. vi.–ix.

her own faith. For many years Flora was to all outward appearance a Mohammedan; but at length, moved by the same spirit of sacrifice and enthusiasm which had stirred Eulogius, and excited by such passages in the Bible as, "Whoso shall deny Me before men, him will I also deny before My Father which is in heaven," she fled from her brother's house—her father was dead—and took refuge among the Christians. The brother, a Mohammedan, searched for her in vain; many priests were thrown into prison on the charge of being accomplices in the abduction; and Flora, unwilling that others should suffer through her fault, returned to her home and confessed herself a Christian. Her brother tried the sternest means at his disposal to compel her to recant, and at last, in a rage at her obstinacy, brought her before the Kādy, or Mohammedan judge, and accused her of apostacy. The child of a Moslem, even though the mother be a Christian, is held in Mohammedan law to be born a Moslem, and apostacy has always been punishable by death. Even now in Turkey the law holds good, though there has been a tacit understanding for the last forty years that it shall not be enforced; and a thousand years ago we must expect to find less tenderness towards renegades. Yet the judge before whom Flora was thus arraigned displayed some compunction towards the unhappy girl. He did not condemn her to death—as he was in law bound to do—or even to imprisonment; he had her severely beaten, and told her brother to take her home and instruct her in the Mohammedan religion. She escaped, however, again, and took refuge with some

Christian friends, and here for the first time she met Eulogius, who conceived for the beautiful and unfortunate young devotee a pure and tender love such as angels might feel for one another. Her mystical exaltation, devout piety, and unconquerable courage, gave her the aspect of a saint in his eyes, and he had not forgotten a detail of this first interview six years later when he wrote to her these words: "Thou didst deign, holy sister, to show me thy neck torn by the scourge, and shorn of the beautiful locks that once hung over it. It was because thou didst regard me as thy spiritual father, and believe me to be pure and chaste as thyself. Softly did I lay my hand on thy wounds; I had it in me to seek to heal them with my lips, had I dared. . . . When I parted from thee I was as one that walketh in a dream, and I sighed without ceasing." Flora and a sister who shared her enthusiasm were removed to a safe place of concealment, and Eulogius did not see her again for some time.

Meanwhile the zeal of the Cordovan Christians was bearing fruit. A foolish priest, Perfectus, had been led into cursing the dominant religion, and had been executed on a great Mohammedan feast-day, when all the world was rejoicing at the termination of the rigorous fast of Ramadan, which had lasted a whole month. The Moslems, men and women, made this feast a special occasion of merry-making, and the execution of the offending priest added a new subject of excitement to the crowds that thronged the streets and sailed on the river and frolicked on the great plain outside the city. The poor priest died bravely, cursing Mohammed and his religion with his last breath, surrounded by a vast crowd of scoffing and pitiless Moslems. The Bishop of Cordova, followed by an army of priests and devotees, took down his body, buried him with the holy relics of St. Acisclus, a martyr of Diocletian's persecution, in whose church he had officiated, and forthwith had him made a saint. The same evening two Moslems were drowned, and this was at once accepted as the judgment of God on the murderers of Perfectus. The black slave, Nasr, who had superintended the execution, died within the year, and the Christians triumphantly declared that Perfectus had predicted his decease: "It was another judgment!"

Soon a monk named Isaac sought an interview with the Kādy, on the pretext of wishing to be converted to the Mohammedan religion; but no sooner had the learned judge explained the doctrines of Islam than the would-be convert turned round, and began to heap maledictions upon the creed which he had asked to be taught. It was no marvel that the astonished Kādy gave him a cuff. "Do you know," said he, "that our law condemns people to death for daring to speak as you have spoken?" "I do," answered the monk; "condemn me to death; I desire it; for I know that the Lord said, 'Blessed are they who are persecuted for righteousness' sake, for theirs is the kingdom of heaven.'" The Kādy was sorry for the man, and begged the Sultan to overlook his crime, but in vain. Isaac was decapitated, and thereupon became a saint, and it was proved conclusively that he had worked many miracles, not only ever since his childhood, but even before he came into the world.

Presently one of the Sultan's guards, Sancho, a pupil of Eulogius, blasphemed Mohammed, and lost his head. Next Sunday six monks rushed before the Kādy and shouted, "We, too, say what our holy brothers Isaac and Sancho said," and forthwith fell to blaspheming Mohammed, and to crying, "Avenge your accursed Prophet! Treat us with all your barbarity!" Their heads were cut off. Three more priests or monks, infected with the fever of suicide, rushed excitedly to present their necks to the headsman. Eleven thus fell in less than two months during the summer of 851.

The great body of the Christians were dismayed at the indiscreet zeal of their brethren. It must not be forgotten that the Spaniards had not so far been remarkable for religious fervour. Their creed sat lightly upon them, and so many of them had been converted to Islam, that the two creeds and the two peoples had become to a considerable extent mixed together in friendly intercourse. The Christians had come to despise their old Latin language and literature; they learned Arabic, and soon were able to write it as well as the Arabs themselves. Eulogius himself deplores this change. The Christians, he says, delight in the Arabic poems and romances instead of the Holy Scriptures and the works of the Fathers. The younger generations know only Arabic; they read the Moslems' books with ardour, form great libraries of them, and find them admirable; while they will not glance at a Christian book. They are forgetting their own language, he adds, and hardly one in a thousand can write a decent Latin letter; yet they indite excellent Arabic verse. The Christians, in fact, found Arab romances and poetry much more entertaining than the writings of the Fathers of the Church. They were growing more and more Arab; more civilized, more refined, and also more indifferent to distinctions of faith. They were grateful to the Moors for treating them well, and the sudden animosity displayed by their excited brethren amazed and shocked them. They endeavoured to avert the threatening storm by showing their brethren the futility of their conduct. They argued with them; reminded them how tolerant the Moslems had always been to the Christians; recalled to them the peaceful teaching of the gospel, and the words of the apostle, that "Slanderers shall not enter the kingdom of heaven;" and told them how the Moslems regarded these deaths with no disquietude, for they argued, "If your religion were true, God would have avenged His martyrs."

These worthy Christians of the common kind, who knew not the force of spiritual exaltation for good and for evil, and only did their duty to their neighbours and said their prayers in the simple, old-fashioned manner, tried in vain to restrain the zealots. They perceived that these continued insults and swift-following punishments must at last end in real persecution. Eulogius, on the contrary, who set himself to answer their objections with texts out of the Bible and the Lives of the Saints, coveted such a result, and the zealots desired nothing better than the fire of persecution. The ecclesiastical authorities, worked upon by the moderate party, and also by the Moorish government, could not permit the spirit of revolt to continue much longer unreproved; the bishops met in council under the presidentship of the Metropolitan of Seville, and though they could not precisely repudiate the former "martyrdoms," since the Church had already canonized the sufferers, yet they ordained that no more exhibitions of the kind should be made, and in furtherance of this decision the leaders of the zealots were thrown into prison. Here Eulogius met Flora again. She had been praying earnestly one day in a church, when she saw beside her a fellow-enthusiast, a sister of that monk Isaac who had been one of the earliest "martyrs." Mary wanted to join her brother in the kingdom of heaven, and Flora resolved to accompany her. They went before the Kādy and did their best to excite his anger by blaspheming the name of Mohammed and his religion. Two young and beautiful girls, professing most sincerely the religion of "peace on earth and goodwill towards men," stood before the magistrate with lips full of cursing and bitterness, reviling his faith as "the work of the devil." But the good judge was not to be roused so easily. He was weary of all this hysterical mania, and had many a time pretended to be deaf when people thrust themselves upon death; he thought it was a pity of these two girls, and wished they would not be so foolish. He would try to induce them to retract, or make as though he had not heard. But they persisted in their heroic purpose, and he had to put them in prison.

Here, in the long confinement, the maidens were daunted, and almost inclined to waver in their sacrificial ardour, when Eulogius came to strengthen and destroy them. His task was the hardest in the world:

to encourage the woman whom he loved with all his soul to go to the scaffold; yet, in spite of every natural and human feeling, this man of iron nerved himself to fan the flame of enthusiasm to the point of martyrdom. It was a daily agony to the unhappy priest, but he never relaxed his efforts in what he believed to be the good cause. He even wrote an entire treatise to convince Flora—who needed it but little—of the supreme beauty and glory of martyrdom for the faith. He spent his days and nights in reading and writing, to banish from his heart those feelings of compunction and love which threatened to shake his resolution. But it was only too firm. Flora and Mary remained constant and undismayed in spite of the anxious efforts of the Kādy to help them to save themselves; and after the final interview, when sentence of death was pronounced, Eulogius saw Flora:—"She seemed to me an angel," he wrote afterwards, glorying in the spiritual triumph. "A celestial illumination surrounded her; her face lightened with happiness; she seemed already to be tasting the joys of the heavenly home. . . . When I heard the words of her sweet mouth, I sought to stablish her in her resolve by showing her the crown that awaited her. I worshipped her; I fell down before this angel, and besought her to remember me in her prayers; and, strengthened by her speech, I returned less sad to my sombre cell." Flora and her companion Mary were executed at last, 24th November, 851, and Eulogius wrote a pæan of joy to celebrate what he deemed a great victory of the Church.

Soon after this, Eulogius and the other priests were

released from prison, and the next year Abd-er-Rahmān II. died, and was succeeded by his son Mohammed, a rigid, cold-hearted egotist, who screwed savings out of the salaries of his ministers, and was universally detested for his meanness and unworthiness. The theologians alone liked him, for he seemed likely to avenge to the full the insults which the excited Christians had poured upon the Mohammedan religion. Churches were demolished, and such severe persecutions were set on foot, that though many Christians had become Moslems when the bishops had officially condemned suicidal martyrdom, many more now followed their example; indeed, according to Eulogius and Alvaro, the majority recanted. The wise and kindly policy of Abd-er-Rahmān and his ministers, who shut their eyes when the Christians were wantonly committing themselves, was now exchanged for a policy of cruel repression, and it is no wonder that apostacy was the rule.

Still, the influence of the little band of zealots was powerful, and had already extended far beyond the limits of Cordova. Toledo made Eulogius its bishop, and when the Sultan refused his consent, the primacy was kept vacant until the zealot should be permitted to occupy it. Two French monks came to Cordova to beg some relics of the holy martyrs, and went back to St. Germain-des-Pres with a handsome bag of bones, which were presently displayed to the faithful at Paris. But a heavy blow was about to fall upon the enthusiasts. Another girl deserted her parents to follow Eulogius; and this time she and her teacher were brought before the Kādy. Eulogius was guilty only of proselytizing, and his legal punishment was but a scourging. But the priest was not made of the stuff that endures the whip. Humble and long-suffering before his God, willing to inflict any torture on his own body for the sake of the faith, he could not submit to be flogged by the infidel. "Make sharp thy sword, judge," he cried; "send my soul to meet my Creator; but think not that I will suffer my body to be lacerated with whips." And here he burst into a flood of maledictions against Mohammed and his religion.

The Kādy would not take upon himself the responsibility of executing the sentence upon so prominent a leader as Eulogius, and the priest was accordingly brought before the privy council. One of the body expostulated with him, and asked why a man of sense and education should voluntarily run his head into peril of death; he could understand fools and maniacs doing so, he said, but Eulogius was of a different stamp. "Listen to me," he added, "I entreat you; yield for once to necessity; retract what you said before the Kādy; say but the word, and you shall go free." But it was too late. Eulogius, though he preferred the position of trainer of martyrs to setting the example himself, could not retreat from his ground with dignity. He must go on to the bitter end. And refusing to retract anything, he was forthwith led out to execution, and died with courage and devotion on March 11, 859.[1]

Deprived of their leader, the Christian martyrs lost heart, and we do not hear of their mad devotion again.

[1] Dozy: livre ii. ch. ix.

VI.

THE GREAT KHALIF.

MY readers may perhaps be disappointed that so far we have but few records of noble deeds or great wars, and that instead of individual heroes we have been chiefly interested in large movements of races and religions. We had, it is true, a stirring outset with Tārik and his Berbers, whose brilliant conquests are no more legendary than is the history of the nineteenth century. We had the great and decisive battle of Tours, but of this the details, which might have proved of surpassing interest, are wanting; and the other engagement with the Franks, the field of Roncesvalles, errs in the opposite direction, for it is overclouded with myth. Since that day, a hundred years have now passed, and we have come to the death of Eulogius and the consequent decline of the Christian martyrs; and in all that century we have been reading of nothing but the struggle between the different races and creeds that made up the mixed population of the Spanish peninsula. But after all, golden deeds are rare, and are too often the invention of poets, whose spiritual minds clothe with the attributes of ideal chivalry what are really the ordinary events of war; while the struggle of race with race and creed with creed is what the world has been incessantly witnessing ever since man came into existence. We must not allow ourselves to think that the history of these large movements is uninteresting because it has not the personal charm of individual acts of heroism. In the devotion of countless unnoticed men and women during the piteous epoch of martyrdom at Cordova there was perhaps more real heroism than in the impetuous deeds of chivalry displayed by rude warriors on the battle-field. It is much easier to be brave in hot blood than to endure the alarms and sufferings of long imprisonment, to look forward with undaunted courage to the day of execution, and keep a firm heart through it all. The Christian martyrs were misguided, they threw away their lives without cause; but their courage is as worthy of admiration as their wisdom is to be pitied. Flora was as real a heroine as if she had sacrificed herself for a worthy sake. Eulogius, with all his bigotry, was of the true hero's mould. And in all these great movements of race or faith there are numberless acts of devotion and fortitude which, though they may escape the eye of the historian, call for as much resolution and endurance as the most brilliant exploits of the soldier. It is often in the little acts of heroism that the hardest duties of mankind are found; and in the conflicts between large bodies of people there are endless opportunities for their exercise.

It is much easier to realize heroic character in a person than in a whole people or even a city; and we are now coming to the career of a man who approached as few have ever done the high ideal of kingly greatness. A great king is the result of a great need. When the nation is sore beset, when the times are full of presage of disaster, and ruin hangs ominously on the horizon ; then the great king comes to rescue his people from danger, to restore order and well-being, and to reign over a realm once more made happy and prosperous by his efforts. The need of such a ruler was anxiously felt at the beginning of the tenth century in Spain. The excited conduct of the Christians of Cordova had been followed by a still more dangerous and widespread rebellion in the provinces. The throne was occupied by incapable sovereigns ; for the energetic policy of Mundhir, who had succeeded his father Mohammed in 886, was arrested by his assassination in 888, and his brother Abdallah, who had instigated the murder, was incapable of dealing courageously with the numerous sources of danger which then menaced the kingdom. His policy was shifty and temporizing ; he alternately tried the effects of force and conciliation, with the usual consequence that both policies failed ; and he was personally so despicable, cruel, and vile, that all parties in his dominions seemed for once to be agreed in their detestation of him, and their resolve to cast off his rule. He had hardly been reigning three years when the greater part of Andalusia was virtually independent. All the various factions of the State were now again in active opposition to the central power. Every nobleman or chief, were he Arab, Berber, or Spaniard, seized the opportunity of a bad and weak sovereign, and general anarchy, to appropriate a portion of the land for his own exclusive

THE GOLDEN TOWER, SEVILLE.

benefit, and from behind his ramparts to defy the Sultan. The old Arab aristocracy, the descendants of the Arab tribes who completed the conquest of Spain, were few and greatly outnumbered by the other races; but though their weakness should have kept them loyal to the Arab kingdom of Cordova, they too turned against it, and established themselves in independent princedoms, especially at Seville, which now became a formidable rival to Cordova. In other cities, though the Arabs were not strong enough to break openly with the Sultan, they gave him but a nominal homage; and the governors of Lorca and Zaragoza were really quite independent of their feeble king. In no place, outside Cordova, where the mercenary guards of the Sultan compelled a certain outward submission, were the Arabs to be counted upon for the defence of the Omeyyad power.

The Berbers were more numerous than the Arabs, and at least equally disaffected. They had abandoned any pretence of submission to the Sultan's authority, and had returned to their old political system of clan government. The western provinces of Spain, such as Estremadura, and the south of Portugal, were now the independent possessions of the Berbers; and they also held various important posts, such as Jaen, in Andalusia itself. The Berber family of Dhu-n-Nun, consisting of the father Mūsa, "a great scoundrel and an abominable thief," and his three sons, who resembled him in their physical strength and their unrivalled brutality, carried fire and sword through the land, and burnt, sacked, and massacred wherever they went.

The Mohammedan Spaniards, who had put on something of Arab civilization along with their new faith, were by no means barbarians like the Berbers; but they were not the less hostile to the central power. The province of Algarve, at the south-west corner of the peninsula, was entirely in their power; and they held numerous independent cities and districts throughout Andalusia. Indeed all the most important cities were in secret or open revolt. Arab governors, Berber chiefs, Spanish renegades, alike joined in repudiating or disregarding the sovereign authority of Abdallah; and most powerful of all, Ibn-Hafsūn, a Christian, who had raised the mountaineers of the province of Elvira (Granada), reigned in perfect security in his rocky fastness, Bobastro, and gave laws to the regions around. Again and again had the Sultan attacked him, and each time suffered defeat; now he was disposed to try the ignominious policy of conciliation, only to find Ibn-Hafsūn quite ready to trick him at that. Murcia, the "land of Theodemir," was independent under a mild and cultivated renegade prince, who governed his subjects wisely, and was beloved by them; who was devoted to poetry, but did not neglect to keep up a considerable army, which included five thousand horsemen. Toledo was, as usual, in revolt, and nothing but the jealousies and divisions of the Christians of the north prevented them from reconquering their long lost territory. Split up as it was into numberless little seigniories, resembling rather the estates or counties of feudal barons than portions of a once powerful realm, Andalusia could have offered

DOOR OF THE MAIDEN'S COURT, ALCAZAR OF SEVILLE.

but an ill-directed resistance to a determined invader.[1]

There were of course some gleams of light amidst all this anarchy. We have said that the province of Murcia was ruled by an enlightened and benevolent prince. The lord of Cazlona was also distinguished for his patronage of poets and the arts; his halls were raised upon marble pillars, and the walls were encrusted with marble and gold; all that makes life enjoyable was to be found within his palace. Ibn-Hajjāj, too, the Arab king—for he was nothing less—of Seville, who had compelled the Sultan to come to terms with him and make him his friend, exercised his unbounded authority in the noblest manner. His city was admirably governed, order reigned there undisturbed, and evil-doers were sternly but justly punished. He kept his state like an emperor; five hundred cavaliers formed his escort, and his royal robe was of brocade, with his name and titles embroidered on it in gold thread. Kings from over the sea sent him presents: silken stuffs from Egypt, learned doctors of the law from Medina, and matchless singers from Baghdad. The beautiful lady "Moon," renowned for her lovely voice, her eloquence, and poetic fire, sang of him thus: "In all the west I find no right noble man save Ibrahim, but he is nobility itself. When one has known the delight of living with him, to dwell in any other land would be misery." The very poets of Cordova were attracted to his brilliant court, where they were sure of a princely welcome. Once only did a poet receive a cold greeting from

[1] Dozy: Hist. des Mus. d'Espagne, livre ii. ch. xi. ff.

Ibrahim the son of Hajjāj. This was one who thought to please the prince by reciting a scurrilous poem on the nobles of Cordova, to whom the ruler of Seville was not well disposed. "You are mistaken," was Ibn-Hajjāj's comment, "if you think that a man like myself can find any gratification in listening to these base calumnies."

Yet these occasional flashes of enlightenment cannot make amends for the general condition of anarchy to which Andalusia had become a prey, by the weakening of the central power, and the aggrandisement of countless petty rulers and brigand chiefs. The country was in a deplorable state, and Cordova itself, now threatened even with conquest at the hands of Ibn-Hafsūn and his bold mountaineers, was given over to mournful sadness. "Without being yet actually besieged, she was already suffering all the ills of beleaguerment." "Cordova," said the Arab historians, "was in the condition of a frontier town exposed to all the attacks of the enemy." Time after time the inhabitants were startled from their sleep, in the midst of night, by the cries of distress raised by the wretched peasants across the river, when the horsemen of Polei were setting the sword to their throats. "The State is menaced with total dissolution," wrote a contemporary witness; "disasters follow one another ceaselessly; thieving and pillaging go on; our wives and children are dragged into slavery." There were universal complaints of the Sultan's want of energy, of his weakness, and his baseness. The troops were grumbling because they were not paid. The provinces had stopped the supplies, and the treasury was empty.

What money the Sultan had been able to borrow, he spent to bribe the few Arabs who still affected to support him in the provinces. The deserted markets showed how trade had been destroyed. Bread had reached a fabulous price. Nobody believed any longer in the future; despair had sunk into all hearts. The bigots, who regarded all public misfortunes as the chastisement of God, and called Ibn-Hafsūn the scourge of the divine wrath, afflicted the city with their doleful prophecies. "Woe to thee, Cordova!" they cried, "woe to thee, sink of defilement and decay, abode of calamity and anguish, thou who hast neither friend nor ally! When the Captain, with his great nose and ugly face, he who is guarded before by Moslems and behind by idolaters — when Ibn-Hafsūn comes before thy gates, then will thy awful fate be accomplished!"

When things were at the worst, a gleam of hope shone upon the miserable inhabitants of the royal city. Abdallah, who was quite as despairing as his subjects, tried for once a bold policy, and in spite of the discouragement of his followers, and the overwhelming numbers of the enemy who surrounded him on every side, he contrived to win a few advantages. Then he did the best thing that he could do for his country: he died on October 15, 912, aged sixty-eight, after a reign of twenty-four unhappy years. His life had seen the fall of the Omeyyad power, a fall sudden and apparently irremediable. The reign of his successor was destined to see as sudden, as complete, a restoration of that power.

The new Sultan was Abd-er-Rahmān III., a grandson of Abdallah. He was only twenty-one when he came to the throne, and there were several uncles and other kinsmen who might be expected to oppose the succession of a mere youth at so troublous a time. Yet no one made any resistance; on the contrary, his accession was hailed with satisfaction on all sides. The young prince had already succeeded in winning the favour of the people and the court. His handsome presence and princely bearing, joined to a singular grace of manner and acknowledged powers of mind, made him generally popular, and it was with a feeling of renewed hope that the Cordovans, who were almost the only subjects he had left, watched the first proceedings of the new Sultan. Abd-er-Rahmān made no attempt to disguise his intentions. He abandoned once and for all the policy of his grandfather, which, in its alternate weakness and cruelty, had worked such injury to the State; and in its place he announced that he would permit no disobedience throughout the dominions of the Omeyyads; he summoned the disaffected nobles and chieftains to submit to his authority; and he let it be clearly understood that he would leave no portion of his kingdom under the control of rebels. The programme was bold enough to satisfy the most sanguine; but there seemed every probability that it would unite all the rebels in all parts in one great league to crush the dauntless young prince. But Abd-er-Rahmān knew his countrymen, and his boldness was well founded. Nearly a generation had passed since Ibn-Hafsūn and the other rebels had raised the standard of insurrection, and every one had come to feel that there had been enough of it. The early zeal that had prompted the Spaniards, Moslem and Christian alike, to strike a blow for their national independence, had now cooled,—such movements never last unless they achieve a complete success at the first white heat of enthusiasm; the leaders were either dead or aged, and a calmer spirit had come over their followers. People had begun to ask themselves what was the good that they had obtained by their fine revolutions? They had not freed Andalusia from the "infidel," but had contrariwise given her over to the worst members of the infidel ranks—to brigand chiefs and adventurers of the vilest stamp. The country was harried from end to end by bands of lawless robbers, who destroyed the tilled fields and vineyards, and turned the land into a howling wilderness. Anything was better than the tyranny of brigandage. The Sultan of Cordova could not make matters worse than they were, and there was a general disposition to see whether he might not possibly improve them.

Consequently, when Abd-er-Rahmān began to lead his army against the rebellious provinces, he found them more than half willing to submit. His troops were inspirited to see their gallant young sovereign at their head—a sight that Abdallah had not permitted them for many years—and they followed him with enthusiasm. The rebels, already tired of their anarchic condition, opened their gates after a mere show of resistance. One after another the great cities of Andalusia admitted the Sultan within their walls. The country to the south of Cordova was the first to submit; then Seville opened her gates; the Berbers of the west were reduced to obedience; and the prince of Algarve hastened to offer tribute. Then the Sultan advanced against the Christians of the province of Regio, where for thirty years the mountain fastnesses had protected the bold subjects of Ibn-Hafsūn, and where no one knew better than Abd-er-Rahmān that no speedy victory was to be won. Yet step by step this difficult region was subdued. Seeing the scrupulous justice and honour of the Sultan, who kept his treaties with the Christians in perfect good faith, and observed the utmost clemency to those who submitted to him, fortress after fortress surrendered. Ibn-Hafsūn himself, in his fastness, remained unconquered and defiant as ever, but he was old, and soon he died, and then it was only a matter of time for the arms of the Sultan to penetrate even into Bobastro. When the Sultan stood at last upon the ramparts of this redoubtable fortress, and looked down from its dizzy heights upon the cliffs and precipices that surrounded the rebel stronghold, he was overcome with emotion, and fell upon his knees to render thanks to God for the great victory.[1] Then he turned to acts of mercy and pardon, and all the days he stayed in the fort he observed a solemn fast. Murcia had now given in its allegiance to the Sultan, and Toledo alone remained unsubdued. The proud city on the Tagus haughtily rejected Abd-er-Rahmān's offer of amnesty, and confidently awaited the siege. But it had to do with a different assailant from the feeble generals who had from time to time reaped disgrace beneath the walls of the Royal City. To prove to its defenders that his

[1] Dozy: Hist. des Mus. d'Espagne, livre ii. ch. xvii.

AQUEDUCT NEAR GRANADA.

siege was no transitory menace, the Sultan quickly built a little town, which he called El-Feth ("Victory"), on the opposite mountain, and there he resided in calm anticipation of the result. Pressed by famine, the city surrendered, and Abd-er-Rahmān III. entered the last seat of rebellion in the dominions which he had inherited from his namesake, the first Abd-er-Rahmān, which now (930) once more reached to their full extent.

It had taken eighteen years to recover the whole breadth of dominion which his predecessors had lost; but the work was done, and the royal power was firmly established over Arabs, Berbers, Spaniards, Moslems and Christians alike. Henceforward Abd-er-Rahmān permitted no special prominence to any party; he kept the old Arab nobility in severe repression; and the Spaniards, who had always been treated by them as base *canaille*, rejoiced to see their oppressors brought low. Henceforth the Sultan was the sole authority in the State; but his authority was just, enlightened, and tolerant. After so many years of confusion and anarchy, the people accepted the new despotism cheerfully. There were no more brigands to destroy their crops and vines; and if the Sultan was absolute in his power, at least he did not abuse it. The country folk returned to the paths of peace and plenty; they were at last free to get rich and to be happy after their own way.

VII.

THE HOLY WAR.

ABD-ER-RAHMĀN III.'s principle of government consisted in retaining the sovereign power entirely in his own hands, and administering the kingdom by officers who owed their elevation wholly to his favour. Above all, he took care to leave no power in the hands of the old Arab aristocracy, who had so ill served previous rulers. The men he appointed to high places were parvenus, people of mean birth, who were the more attached to their master because they knew that but for him they would be trampled upon by the old Arab families. The force he employed to sustain the central power was a large standing army, at the head of which stood his select body-guard of *Slavs*, or purchased foreigners. They were originally composed chiefly of men of Slavonian nationality, but came by degrees to include Franks, Galicians, Lombards, and all sorts of people, who were brought to Spain by Greek and Venetian traders, and sold while still children to the Sultan, to be educated as Moslems. Many of them were highly cultivated men, and naturally attached to their master. They resembled in many respects the corps of Mamlūks which Saladin's successors introduced into Egypt as a body-guard, and which subsequently attained such renown as Sultans of Egypt and Syria. Like that body of purchased Turkish and Circassian slaves, they had their own slaves under them, were granted estates by the Sultan, and formed a sort of feudal retainers, prepared to serve their lord at the head of their own followers whenever he might call upon them. Like the Egyptian Mamlūks, too, they came after a while to such a pitch of influence that they took advantage of the decay of the central power, which followed upon the death of Abd-er-Rahmān III. and his successor, to found independent dynasties for themselves, and thus contribute to the final overthrow of the Moslem domination in Spain.

With the aid of his "Slavs," the Sultan not only banished brigandage and rebellion from Spain, but waged war with the Christians of the north with brilliant success. The Mohammedan realm was menaced by more dangers than those of internal anarchy. It was pressed between two threatening and warlike kingdoms, each of which required to be kept in watchful check. To the south the newly founded empire of the Fātimite Khalifs in North Africa was a standing menace. It was natural that the rulers of the Barbary coast should remember that the Arabs before them had used Africa as a stepping stone to Spain ; the traditional policy of the African dynasties was to compass, if possible, the annexation of the fair provinces of Andalusia. It was only by skilfully working upon the sectarian schisms, and consequent insurrections, which divided the Berbers of Africa, that the Sultan succeeded in keeping the

Fātimites at a distance. He did succeed, however, so well, that at one time a great part of the Barbary coast paid homage to the ruler of Spain, who also obtained possession of the important fortress of Ceuta. A great part of the Spanish revenue was devoted to building a magnificent fleet, with which Abd-er-Rahmān disputed with the Fātimites the command of the Mediterranean.

On the opposite side, on the north, the Moslem power had to deal with an even more threatening enemy. The Christians of the Asturias had sprung from very small beginnings, but they were now increasing in strength, and they had the stimulating thought to spur them on, that they were reconquering their own land. When first they had felt the shock of the Moslem invasion, their rout had been utter and complete. They had fled to the mountains of the Asturias, where their trifling numbers and the inaccessibility of their situation gave them safety from the Mohammedan attack. Pelagius, the "old Pelayo" of the ballad, had but thirty men and ten women with him in the cave of Covadonga, which became the refuge of the Gothic Christians; and the Arabs did not think it worth while to hunt down the little remnant of refugees. Here, in the recesses of the cave, which was approached through a long and narrow mountain pass, and entered by a ladder of ninety steps, a handful of men might have set an army at defiance.

The Arab historian [1] thus contemptuously describes the origin of the Christian kingdom: "During

[1] Ibn-Hayyān, in Makkary, ii. 34.

Anbasa's administration a despicable barbarian, whose name was Pelayo, rose in the land of Galicia, and, having reproached his countrymen for their ignominious dependence and their cowardly flight, began to stir them up to avenge their past injuries and to expel the Moslems from the land of their fathers. From that moment the Christians of Andalus began to resist the attacks of the Moslems on such districts as had remained in their possession, and to defend their wives and daughters. The commencement of the rebellion happened thus: there remained no city, town, or village in Galicia but what was in the hands of the Moslems, with the exception of a steep mountain on which this Pelayo took refuge with a handful of men; there his followers went on dying through hunger, until he saw their numbers reduced to about thirty men and ten women, having no other food for support than the honey which they gathered in the crevices of the rock which they themselves inhabited like so many bees. However, Pelayo and his men fortified themselves by degrees in the passes of the mountain, until the Moslems were made acquainted with their preparations; but, perceiving how few they were, they heeded not the advice conveyed to them, and allowed them to gather strength, saying, 'What are thirty barbarians, perched up on a rock? They must inevitably die!'" "Would to God!" adds another historian—"Would to God that the Moslems had then extinguished at once the sparks of a fire which was destined to consume the whole dominions of Islam in those parts!"

The little band of refugees was strengthened from

time to time by fresh accessions, and, by degrees waxing more confident, came forth from their stronghold, and began to harass the Berbers who formed the frontier settlers. The Moors were at length compelled to seek out the intrepid raiders in their cavern; but the result was discouraging; they were driven back pell-mell with great loss. In 751 Alfonso of Cantabria (where the Moslems had never penetrated), having married the daughter of Pelayo and thus united the Christian forces, roused the northern provinces against the Moors, and, joined by the Galicians of the west, began a series of brilliant campaigns, by which the enemy was driven step by step further south. One after the other the cities of Braga, Porto, Astorga, Leon, Zamora, Ledesma, Salamanca, Saldaña, Segovia, Avila, Osma, Miranda, were recovered from the Moslems, and the Christian frontier was now pushed as far as the great Sierra, and Coimbra, Coria, Talavera, Toledo, Guadalaxara, Tudela, and Pamplona became the Moslem border fortresses. Alfonso had in fact recovered the provinces of Old Castile, Leon, Asturias, and Galicia; but the scanty band of Christians had neither money nor serfs wherewith to build fortifications and cultivate the fields over so immense an area: they contented themselves with leaving the conquered country as a debatable land between them and the Moors, and retired to the districts bordering the Bay of Biscay until such time as their numbers should justify the occupation of a wider area.

In the ninth century they were in a position to advance upon the territory they had already in part recovered from the Moors. They spread over Leon, and built the fortresses of Zamora, San Estevan de Gormaz, Osma, and Simancas, to overawe the enemy. The debatable land was now much narrower, and the hostile forces were almost in contact at various places along the frontier. At the beginning of the tenth century the Moors of the borders made a strenuous effort to regain their lost dominions; but the Christians, aided by the men of Toledo, and by Sancho, King of Navarre, who had become the bulwark of Christianity in the north, defeated them severely, and began to harry the country over the border. The forays of the Christians were a terrible curse to their victims; they were rude, unlettered people, and few of them could even read; their manners were on a par with their education; and their fanaticism and cruelty were what might be expected from such uncouth barbarians. Seldom did the soldiery of Leon give quarter to a defenceless foe, and we may look in vain for the fine chivalry and toleration of the Arabs; where the latter spared nobly, the rough robbers of Leon and Castile massacred whole garrisons, cities full of inhabitants, and those whom they did not slaughter they made slaves.

Abd-er-Rahmān III. had hardly been seated two years on the throne when Ordoño II. of Leon carried a devastating foray to the walls of Merida; and so affrighted were the people of Badajoz that they hastened to conciliate him with blackmail. These cities are not very far from Cordova; only the lofty heights of the Sierra Morena separated the capital of the Omeyyads from the companies of Ordoño. The situation was fraught with danger. The young Sultan, had he been a coward, might have excused himself from instant action on the plea that Merida had not yet recognized his authority, and that it was not his affair if the Christians harried rebellious provinces. This, however, was not Abd-er-Rahmān's policy or temper. He collected his troops and sent an expedition to the north, which made a successful raid into the Christian territories; and the following year, 917, he ordered a second attack. This was defeated with heavy loss by Ordoño before the walls of San Estevar de Gormaz, and the brave Arab general, seeing that the fight was lost, threw himself among the enemy, and died sword in hand. The King of Leon had the pitiful cowardice to nail the head of this gallant soldier to the gate of the fortress, side by side with that of a pig. Encouraged by this success, the armies of Leon and Navarre ravaged the country about Tudela in the following year, but not with equal impunity, for they were twice beaten by the Cordovan troops. Seeing, however, that it took a good deal of defeat to daunt the Christians, Abd-er-Rahmān resolved upon stronger measures. In 920 he took command of the army himself, and by rapid marches and skilful strategy surprised Osma, and razed the fortress to the ground; destroyed San Estevan, which he found deserted by its garrison; and then turned towards Navarre. Twice did he drive Sancho from the field, and when the forces of Navarre were reinforced by those of Leon, and the Christians had the best of the natural position, the Sultan delivered battle with them in the Val de Junqueras (Vale of Reeds), and totally routed their combined array. Incensed by the obstinate defence of the borderers, the Moslems put the garrison of Muez to the sword; and it is unfortunately true that in some of these campaigns the Moors imitated the barbarities of their antagonists, especially when their armies included a considerable admixture of African troops, who were notoriously savage.

Nothing could exceed the heroic determination of the defeated Christians; barbarous they were, but they had the courage of men: routed again and again, they ever rose with fresh heart from the disaster. The very year after the fatal battle in the Valley of Reeds, Ordoño, who was the soul of the Christian resistance, led his men on another raid over the borders; and in 923 Sancho of Navarre, not to be behindhand, recaptured some strong castles. Thus roused once more, the Sultan set out for the north, filled with a stern resolve; he sacked and burned all that came in his way; the cities emptied as he approached, so terrible was the dread he inspired; and he entered the deserted capital of Pamplona, driving Sancho away in confusion as he approached. The cathedral and many of the houses of the capital were ruthlessly destroyed, and Navarre was at his feet. About the same time Ordoño of Leon died, and the civil war which arose between his sons gave the Sultan time to attend to other matters.

On his return from this triumphant campaign, Abd-er-Rahmân III. assumed a new title. Hitherto the rulers of Andalusia had contented themselves with such titles as *Emīr* (governor), *Sultan* (dominator), "son of the Khalifs." Although they were the heirs of the Omeyyad Khalifs, and never recognized the Abbāsides who had overturned them, the Andalusian Sultans had not hitherto asserted their claim to the spiritual title: they had considered that the name of Khalif should not be held by those who had no authority over the Holy Cities of Islam, Mekka and Medina, and had been content to leave the Abbāsides in undisputed possession of the name. Now, however, when it was known in Spain that the Abbāside Khalifs no longer exercised any real authority outside the city of Baghdad, and were little better than prisoners even there, in consequence of the growing independence of the various local dynasties, Abd-er-Rahmān, in 929, assumed his title of Khalif with the style of *En-Nāsir li-dīni-llāh*, "The Defender of the Faith of God."[1]

The Khalif had still thirty years more to reign when he adopted this new name; and they were filled chiefly with wise and cultivated administration at home, and with constant, even annual, expeditions against the Christians, against whom he was indeed a "Defender" of his religion. The civil war, which had for a time neutralized the power of the Leonese, had now given place to the authority of a worthy successor of the great Ordoño. Ramiro II. succeeded in 931, and his warlike character soon asserted itself in resolute opposition to the Khalif's armies. Not long afterwards a formidable league was formed in the north between the Christians and the Arab governor of Zaragoza, and Abd-er-Rahmān hastened to demolish

[1] Dozy, livre iii.

the coalition. In 937 he reduced Zaragoza, and, marching on Navarre, spread such terror around his way that the Queen Regent, Theuda, hastily paid him homage as her suzerain. Ramiro, however, was no party to this surrender. He gathered his men together, and inflicted a tremendous defeat on the Moslems in 939 at Alhandega. Fifty thousand Moors fell upon the field: the Khalif himself barely escaped with his life, and found himself flying through the country with less than fifty horsemen. That disastrous year was long known in Andalusia as the "Year of Alhandega."

Had the Christians pressed their advantage, a different history of Spain would perhaps have had to be written; but, as usual, internecine jealousies among the Christian princes came to the help of the Khalif, and while his foes quarrelled among themselves he repaired his disaster, recruited his army, and made ready for another campaign. The civil war which thus aided him had its origin in the revolt of Castile from the Leonese supremacy. The Count of Castile at this time was the celebrated Fernando Gonzalez, of whom many minstrels have sung. He is one of the great Spanish heroes, and was mated to a heroine. Twice did his wife rescue him from the prison into which he had been cast by his jealous neighbours of Navarre and Leon, and the second time she did it by exchanging clothes with her husband and exposing herself to the fury of his jailers. The earlier occasion was before their marriage, when he was on his way to her father Garcia's court at Navarre, to ask her hand in marriage, and the perfidious king laid hands upon him. A ballad tells the story of his release:

> They have carried afar into Navarre the great Count of Castille,
> And they have bound him sorely, they have bound him hand and heel. . . .
> And there is joy and feasting because that lord is ta'en,
> King Garci in his dungeon holds the doughtiest lord in Spain.

The poet goes on to tell how a Norman knight was riding through Navarre—

> For Christ his hope he came to cope with the Moorish scimitar;

and how he told Garcia's daughter of the captivity of Gonzalez, and how grievous an injury it was to the cause of Christian Spain—

> The Moors may well be joyful, but great should be our grief,
> For Spain has lost her guardian, when Castile has lost her chief;
> The Moorish host is pouring like a river o'er the land—
> Curse on the Christian fetters that bind Gonzalez' hand!

And the Norman knight prayed the princess to set the prisoner free.

> The lady answered little, but at the mirk of night,
> When all her maids are sleeping, she hath risen and ta'en her flight;
> She hath tempted the Alcayde with her jewels and her gold,
> And unto her his prisoner that jailor false hath sold.[1]

So the princess took the Count out of his dungeon, and together they rode to Castile.

At the time we have now reached, this is an old story, for Gonzalez had been married many a year, and had determined that Castile should be a separate kingdom, no longer under the suzerainty of Leon. For this he was again captured and imprisoned by Ramiro, and only released when it was apparent that the people of Castile would have no other lord but him, and would even pay their homage to a mere

[1] Lockhart: Spanish Ballads.

statue of their Count sooner than recognize a Leonese governor. Then the king let him out, after making him swear to remain subject to the kingdom of Leon and to give his daughter in marriage to Ordoño the son of Ramiro. After this humiliation, Fernando Gonzalez was less eager to fight beside the men of Leon against the Moors; he resolved to let the Leonese take their share of humiliation. But this was not to be in the days of the great Ramiro; for he won another victory over the Moslems, near Talavera, in 950, and the next year he died in undiminished glory.

On his death, Gonzalez began to play the part of king-maker. He espoused the cause of Sancho against his brother, Ordoño III., and when Sancho succeeded the latter, in 957, Gonzalez turned about and expelled the new king from Leon, and set up a wretched cripple, Ordoño IV., surnamed the Wicked, in his stead. Sancho took refuge with his grandmother, Theuda, the Queen of Navarre, and they presently appealed to the Khalif of Cordova to help them in their difficulties. Sancho was a martyr to corpulency; he could not even walk without being held up. He resolved to consult the eminent doctors of Cordova, whose skill was famous over all the world. So Queen Theuda sent ambassadors to Abd-er-Rahmān, who in return despatched the great Jewish physician, Hasdai, to undertake the cure of Sancho the Fat. But he laid down certain conditions, among which was the surrender of a number of castles, and the personal appearance of Sancho and the Queen Theuda at Cordova. It was a hard thing to make the long journey to the Moorish Court, and to feel that she was there as a sort of show, in witness to the Khalif's power ; but the Queen went, with her son, the King of Navarre, and her grandson, the exiled King of Leon. Abd-er-Rahmān received them with all the gorgeous ceremony and all the native courtesy which belonged to him ; and not only did Sancho speedily get rid of his fatness under the care of Hasdai, but he returned to the north, supported by the armies of the Khalif, who restored him to the throne of Leon in 960.

In the following year the great Khalif died. He was seventy years old, and his reign, of nearly fifty, had brought about such a change in the condition of Spain as the wildest imagination could hardly conjure up. When he came to the throne, a youth of twenty-one, his inheritance was the prey to a thousand brigand chiefs or local adventurers ; the provinces had set up their own rulers ; the many factions into which the population was divided had each and all defied the authority of the Sultan ; and anarchy and plunder devastated the land. On the south the African dynasty of the Fātimites threatened to engulf Spain in their empire ; on the north the Christian princes seemed ready to descend upon their ancestral dominions and drive the Moors from the land. Out of this chaos and vision of imminent destruction Abd-er-Rahmān had evolved order and prosperity. Before half his reign was over he had restored peace and good government throughout the length and breadth of the Moslem dominions ; he had banished the authority of parties, and established the absolute power of the Sultan over all classes of his subjects. In the second half he maintained the dignity and might of his State against outside foes; held the African despots at a distance, planted a garrison at Ceuta to withstand their advance, and contended with them on equal terms on the sea; and in the north he curbed the growing power of the Christians of Leon, Castile, and Navarre, and so convinced them of his superiority that they even came to him to settle their differences and restore them to their rights. He had rescued Andalusia both from herself and from subjection by the foreigner.[1] And he had not only saved her from destruction; he had made her great and happy. Never was Cordova so rich and prosperous as under his rule; never was Andalusia so well cultivated, so teeming with the gifts of nature, brought to perfection by the skill and industry of man; never was the State so triumphant over disorder, or the power of the law more widely felt and respected. Ambassadors came to pay him court from the Emperor of Constantinople, from the kings of France, of Germany, of Italy. His power, wisdom, and opulence, were a byword over Europe and Africa, and had even reached to the furthest limits of the Moslem empire in Asia. And this wonderful change had been wrought by one man, with everything against him: the restoration of Andalusia from the hopeless depths of misery to the height of power and prosperity had been effected by the intellect and will alone of the Great Khalif Abd-er-Rahmān III.

The Moorish historians describe this resolute man

[1] Dozy: Hist. des Mus. d'Espagne, livre iii. p. 90.

in colours that seem hardly consistent with his strong imperious policy: nevertheless, they describe him faithfully as "the mildest and most enlightened sovereign that ever ruled a country. His meekness, his generosity, and his love of justice became proverbial. None of his ancestors ever surpassed him in courage in the field and zeal for religion; he was fond of science, and the patron of the learned, with whom he loved to converse." Many anecdotes are told of his strict justice and impartiality.

The Arab historian tells us that after his death a paper was found in the Khalif's own handwriting, in which he had carefully noted those days in his long reign which had been free from all sorrow; they numbered only fourteen. "O man of understanding, wonder and observe how small a portion of unclouded happiness the world can give even to the most fortunate!" [1]

[1] Makkary: Hist. Moh. Dynast. ii. 146, 147.

VIII.

THE CITY OF THE KHALIF.

"CORDOVA," says an old Arab writer, "is the Bride of Andalusia. To her belong all the beauty and the ornaments that delight the eye or dazzle the sight. Her long line of Sultans form her crown of glory; her necklace is strung with the pearls which her poets have gathered from the ocean of language; her dress is of the banners of learning, well knit together by her men of science; and the masters of every art and industry are the hem of her garments." So did the Oriental historian clothe the city he loved with the far-fetched imagery of the East. Cordova, under the rule of the Great Khalif, was indeed a capital to be proud of; and except perhaps Byzantium, no city of Europe could compare with her in the beauty of her buildings, the luxury and refinement of her life, and the learning and accomplishments of her inhabitants. When we remember that the sketch we are about to extract from the records of Arabian writers, concerning the glories of Cordova, relate to the tenth century, when our Saxon ancestors dwelt in wooden hovels and trod upon dirty straw, when our language was unformed, and such accomplishments as reading and writing were almost confined to a few monks, we can to some extent realize the extraordinary civilization of the Moors. And when it is further recollected that all Europe was then plunged in barbaric ignorance and savage manners, and that only where the remnants of the Roman Empire were still able to maintain some trace of its ancient civilization, only in Constantinople and some parts of Italy, were there any traces of refinement, the wonderful contrast afforded by the capital of Andalusia will be better appreciated.

Another Arab writer says that Cordova "is a fortified town, surrounded by massive and lofty stone walls, and has very fine streets. It was in times of old the residence of many infidel kings, whose palaces are still visible within the precincts of the walls. The inhabitants are famous for their courteous and polished manners, their superior intelligence, their exquisite taste and magnificence in their meals, dress, and horses. There thou wouldst see doctors shining with all sorts of learning, lords distinguished by their virtues and generosity, warriors renowned for their expeditions into the country of the infidels, and officers experienced in all kinds of warfare. To Cordova came from all parts of the world students eager to cultivate poetry, to study the sciences, or to be instructed in divinity or law; so that it became the meeting-place of the eminent in all matters, the abode of the learned, and the place of resort for the studious; its interior was always filled with the eminent and the noble of all countries, its literary men and soldiers were continually vying with each other to gain renown, and its precincts never ceased to be the arena of the distinguished, the racecourse of readers, the halting-place of the noble, and the repository of the true and virtuous. Cordova was to Andalus what the head is to the body, or what the breast is to the lion." [1]

Oriental praise is apt to be somewhat high flown; but Cordova really deserved the praise that has been lavished upon it. In its present state it is impossible to form any conception of the extent and beauty of the old Moorish capital in the days of the Great Khalif. Its narrow streets of whitewashed houses convey but a faint impression of its once magnificent extent; the palace, Alcazar, is in decay, and its ruins are used for the vile purpose of a prison; the bridge still spans the Guadalquivir, however, and the noble mosque of the first Omeyyad is still the wonder and delight of travellers. But in the time of Abd-er-Rahmān III., or perhaps a little later, when a great minister added a new faubourg, it was at its best. Historians are divided as to its extent, but a length of at least ten miles seems to be the most probable dimension. The banks of the Guadalquivir were bright with marble houses, mosques, and gardens, in which the rarest flowers and trees of other countries were carefully cultivated, and the Arabs introduced their system of irrigation, which the Spaniards, both before and since, have never equalled. The first Omeyyad Sultan imported a date tree from Syria, to remind him of his old home; and to it he dedicated a sad little poem to bewail his exile. It was planted in the garden which he had laid out in imitation of that of his grandfather Hishām at Damascus, where he had played as a child. He sent agents all over the world to bring him the rarest

[1] Makkary, i. book iii.

exotics, trees, plants, and seeds; and so skilful were the Sultan's gardeners that these foreign importations were speedily naturalized, and spread from the palace over all the land. The pomegranate was thus introduced by means of a specimen brought from Damascus. The water by which these numerous gardens were supplied was brought from the mountains (where vestiges of hydraulic works may still be seen) by means of leaden pipes, through which it was conducted to numerous basins, some of gold or silver, others of inlaid brass, and to lakes, reservoirs, tanks, and fountains of Grecian marble.

The historians tell us marvellous things about the Sultan's palaces, with their splendid gates, opening upon the gardens or the river, or again giving entrance to the Great Mosque, whither the Sultan betook himself on Fridays, over a path covered from end to end with rich carpets. One of these palaces was called the Palace of Flowers, another the Palace of Lovers, a third the Palace of Contentment, and another the Palace of the Diadem, and so forth, while one retained the name of the old home of the Omeyyads, and was called "Damascus." Its roofs rested upon marble columns, and its floors were inlaid with mosaics; and so beautiful was it, that a poet sang, "All palaces in the world are nothing when compared to Damascus, for not only has it gardens with the most delicious fruits and sweet-smelling flowers, beautiful prospects and limpid running waters, clouds pregnant with aromatic dew, and lofty buildings; but its night is always perfumed, for morning pours on it her grey amber, and night her black musk." Some of the gardens of Cordova had

EXTERIOR OF THE GREAT MOSQUE AT CORDOVA.

tempting names, which seem to invite one to repose beside the trickling waters and enjoy the sweet scent of the flowers and fruit. The "Garden of the Water-wheel" gives one a sense of lazy enjoyment, listening to the monotonous creaking of the wheel that pumped up the water to the level of the garden beds; and the "Meadow of Murmuring Waters" must have been an entrancing spot for the people of Cordova in the hot weather. The quiet flow of the Guadalquivir was a constant delight to the inhabitants; for the Eastern (and the Moors of Spain were Easterns in everything but longitude) loves nothing better than a view over a rippling stream. It was spanned by a noble bridge of seventeen arches, which still testifies to the engineering powers of the Arabs. The whole city was full of noble buildings, among which were counted more than fifty thousand houses of the aristocracy and official classes, more than a hundred thousand dwellings for the common people, seven hundred mosques, and nine hundred public baths. The last were an important feature in all Moslem towns, for among the Mohammedans cleanliness is not "next to godliness," but is an essential preparation for any act of prayer or devotion. While the mediæval Christians forbade washing as a heathen custom, and the monks and nuns boasted of their filthiness, insomuch that a lady saint recorded with pride the fact that up to the age of sixty she had never washed any part of her body, except the tips of her fingers when she was going to take the Mass — while dirt was the characteristic of Christian sanctity, the Moslems were careful in the most minute particulars of cleanliness, and dared not approach their God until their bodies were purified. When Spain had at last been restored to Christian rulers, Philip II., the husband of our English Queen Mary, ordered the destruction of all public baths, on the ground that they were relics of infidelity.

Among the great architectural beauties of Cordova, the principal mosque held, and still holds, the first place. It was begun in 784 by the first Abd-er-Rahmān, who spent 80,000 pieces of gold upon it, which he got from the spoils of the Goths. Hishām, his pious son, completed it, in 793, with the proceeds of the sacking of Narbonne. Each succeeding Sultan added some new beauty to the building, which is one of the finest examples of early Saracenic art in the world. One put the gold on the columns and walls; another added a new minaret; another built a fresh arcade to hold the swelling congregations. Nineteen is the number of the arcades from east to west, and thirty-one from north to south; twenty-one doors encrusted with shining brass admitted the worshippers; 1,293 columns support the roof, and the sanctuary was paved with silver and inlaid with rich mosaics, and its clustered columns were carved and inlaid with gold and lapis-lazuli. The pulpit was constructed of ivory and choice woods, in 36,000 separate panels, many of which were encrusted with precious stones and fastened with gold nails. Four fountains for washing before prayer, supplied with water from the mountains, ran night and day; and houses were built at the west side of the mosque, where poor travellers and homeless people were

GATE OF THE MOSQUE OF CORDOVA.

hospitably entertained. Hundreds of brass lanterns, made out of Christian bells, illumined the mosque at night, and a great wax taper, weighing fifty pounds, burnt night and day at the side of the preacher during the month of fasting. Three hundred attendants burnt sweet-smelling ambergris and aloes wood in the censers, and prepared the scented oil which fed the ten thousand wicks of the lanterns. Much of the beauty of this mosque still remains. Travellers stand amazed among the forest of columns, which open out in apparently endless vistas on all sides. The porphyry, jasper, and marbles are still in their places; the splendid glass mosaics, which artists from Byzantium came to make, still sparkle like jewels on the walls; the daring architecture of the sanctuary, with its fantastic crossed arches, is still as imposing as ever; the courtyard is still leafy with the orange-trees that prolong the vistas of columns. As one stands before the loveliness of the Great Mosque, the thought goes back to the days of the glories of Cordova, the palmy days of the Great Khalif, which never will return.

Even more wonderful, though not more beautiful, was the city and palace of Ez-Zahrā, which Abd-er-Rahmān III. built as a suburb to Cordova. One of his wives, whose name was Ez-Zahrā, "the Fairest," to whom he was devotedly attached, once begged him to build her a city which should be called after her name. The Great Khalif, like most Mohammedan sovereigns, delighted in building, and he adopted the suggestion. He at once began to found a city at the foot of the mountain called the "Hill of the

Bride," over against Cordova, and a few miles distant. Every year he spent a third of his revenues upon this building; and it went on all the twenty-five remaining years of his reign, and fifteen years of the reign of his son, who made many additions to it. Ten thousand workmen laboured daily at the task, and six thousand blocks of stone were cut and polished every day for the construction of the houses of the new city. Some three thousand beasts of burden were daily used to carry the materials to the spot, and four thousand columns were set up, many of which were presents from the Emperor of Constantinople, or came from Rome, Carthage, Sfax, and other places, besides the home marbles quarried at Tarragona and Almeria. There were fifteen thousand doors, coated with iron or polished brass. The Hall of the Khalifs at the new city had a roof and walls of marble and gold, and in it was a wonderful sculptured fountain, a present from the Greek Emperor, who also sent the Khalif a unique pearl. In the midst of the hall was a basin of quicksilver; at either side were eight doors set in ivory and ebony, and adorned with precious stones. When the sun shone through these doors, and the quicksilver lake was set quivering, the whole room was filled with flashes like lightning, and the courtiers would cover their dazzled eyes.

The Arabian authors delight in telling of the wonders of this "City of the Fairest," Medinat-Ez-Zahrā, as it was called, after the Khalif's mistress. "We might go to a great length were we only to enumerate all the beauties, natural as well as artificial, contained within the precincts of Ez-Zahrā," writes one: "the running streams, the limpid waters, the luxuriant gardens, the stately buildings for the household guards, the magnificent palaces for the high functionaries of State; the throng of soldiers, pages, and slaves, of all nations and religions, sumptuously attired in robes of silk and brocade, moving to and fro through its broad streets; or the crowd of judges, theologians, and poets, walking with becoming gravity through the magnificent halls and ample courts of the palace. The number of male servants in the palace has been estimated at thirteen thousand seven hundred and fifty, to whom the daily allowance of flesh meat, exclusive of fowls and fish, was thirteen thousand pounds; the number of women of various kinds and classes, comprising the harīm of the Khalif, or waiting upon them, is said to have amounted to six thousand three hundred and fourteen. The Slav pages and eunuchs were three thousand three hundred and fifty, to whom thirteen thousand pounds of flesh meat were distributed daily, some receiving ten pounds each, and some less, according to their rank and station, exclusive of fowls, partridges, and birds of other sorts, game and fish. The daily allowance of bread for the fish in the pond of Ez-Zahrā was twelve thousand loaves, besides six measures of black pulse which were every day macerated in the waters. These and other particulars may be found at full length in the histories of the times, and recorded by orators and poets who have exhausted the mines of eloquence in their description; all who saw it owned that nothing similar to it could be found in the territories of Islam. Travellers from distant lands, men of all ranks and professions in life, following various religions,—princes, ambassadors, merchants, pilgrims, theologians, and poets—all agreed that they had never seen in the course of their travels anything that could be compared to it. Indeed, had this palace possessed nothing more than the terrace of polished marble overhanging the matchless gardens, with the golden hall and the circular pavilion, and the works of art of every sort and description—had it nothing else to boast of but the masterly workmanship of the structure, the boldness of the design, the beauty of the proportions, the elegance of the ornaments, hangings, and decorations, whether of shining marble or glittering gold, the columns that seemed from their symmetry and smoothness as if they had been turned by lathes, the paintings that resembled the choicest landscapes, the artificial lake so solidly constructed, the cistern perpetually filled with clear and limpid water, and the amazing fountains, with figures of living beings—no imagination however fertile could have formed an idea of it. Praise be to God Most High for allowing His humble creatures to design and build such enchanting palaces as this, and who permitted them to inhabit them as a sort of recompense in this world, and in order that the faithful might be encouraged to follow the path of virtue, by the reflection that, delightful as were these pleasures, they were still far below those reserved for the true believer in the celestial Paradise!"

In the palace of Ez-Zahrā the Khalif received the Queen of Navarre and Sancho, and gave audience to great persons of State. Here he sat to welcome the ambassadors which the Greek Emperor sent to his court at Cordova:

"Having appointed Saturday the eleventh of the month of Rabi' el-Awwal, of the year 338 [A.D. 949], and fixed upon the vaulted hall in his palace of Ez-Zahrā as the place where he would receive their credentials, orders were issued to the high functionaries of State and to the commanders of the forces to prepare for the ceremony. The hall was beautifully decorated, and a throne glittering with gold and sparkling with gems was raised in the midst. On either hand of the throne stood the Khalif's sons; next to them the vizirs, each in his post to the right and left; then came the chamberlains, the sons of vizirs, the freedmen of the Khalif, and the officers of the household. The court of the palace was strewn with the richest carpets and most costly rugs, and silk awnings of the most gorgeous kind were thrown over the doors and arches. Presently the ambassadors entered the hall, and were struck with astonishment and awe at the magnificence displayed before them and the power of the Sultan before whom they stood. Then they advanced a few steps, and presented a letter of their master, Constantine, son of Leo, Lord of Constantinople, written in Greek upon blue paper in golden characters."

Abd-er-Rahmān had ordered the most eloquent orator of the court to make a suitable speech upon the occasion; but hardly had he begun to speak, when the splendour of the scene, and the solemn silence of the great ones there assembled, so overawed him, that his tongue clove to the roof of his mouth, and he fell senseless on the floor. A second essayed to fill his place, but he had not got very far in his address when he too suddenly broke down.

So interested was the Great Khalif in building his new palace that he omitted to go to the mosque for three successive Fridays; and when at last he made his appearance, the preacher threatened him with the pains of hell for his negligence.

Beautiful as were the palaces and gardens of Cordova, her claims to admiration in higher matters were no less strong. The mind was as lovely as the body. Her professors and teachers made her the centre of European culture; students would come from all parts of Europe to study under her famous doctors, and even the nun Hroswitha, far away in her Saxon convent of Gaudersheim, when she told of the martyrdom of St. Eulogius, could not refrain from singing the praises of Cordova, "the brightest splendour of the world." Every branch of science was seriously studied there, and medicine received more and greater additions by the discoveries of the doctors and surgeons of Andalusia than it had gained during all the centuries that had elapsed since the days of Galen. Albucasis (or Abu-l-Kāsim Khalaf, to give him his proper name) was a notable surgeon of the eleventh century, and some of his operations coincided with the present practice. Avenzoar (Ibn Zohr) a little later made numerous important medical and surgical discoveries. Ibn Beytar, the botanist, travelled all over the East to find medicinal herbs, on which he wrote an exhaustive treatise; and Averroes, the philosopher, formed the chief link in the chain

HISPANO-MORESCO VASE. (*Preserved at Granada.*)

which connects the philosophy of ancient Greece with that of mediæval Europe. Astronomy, geography, chemistry, natural history—all were studied with ardour at Cordova ; and as for the graces of literature, there never was a time in Europe when poetry became so much the speech of everybody, when people of all ranks composed those Arabic verses which perhaps suggested models for the ballads and canzonettes of the Spanish minstrels and the troubadours of Provence and Italy. No speech or address was complete without some scrap of verse, improvised on the spur of the moment by the speaker, or quoted by memory from some famous poet. The whole Moslem world seemed given over to the Muses ; Khalifs and boatmen turned verses, and sang of the loveliness of the cities of Andalusia, the murmur of her rivers, the beautiful nights beneath her tranquil stars, and the delights of love and wine, of jovial company and stolen meetings with the lady whose curving eyebrows had bewitched the singer.

In the arts Andalusia was pre-eminent ; such buildings as the "City of the Fairest," or the mosque of Cordova, could not have been erected unless her workmen had been highly skilled in their handicrafts. Silk weaving was among the most cherished arts of Andalusia ; it is said that there were no less than one hundred and thirty thousand weavers in Cordova alone ; but Almeria had the greatest name for her silks and carpets. Pottery was carried to great perfection, and it was from the island of Majorca, where the potters had attained to the art of producing a ware shining with iridescent gold or copper lustre, that the Italian pottery obtained its name of Majolica. Glass vessels, as well as others of brass and iron, were made at Almeria, and there are some beautiful specimens of delicate ivory carvings still in existence, which bear the names of great officers of the court of Cordova. These arts were no doubt imported from the East, but the Moorish workmen became apt pupils of their Byzantine, Persian, and Egyptian masters. In jewellery an interesting relic of the son of the Great Khalif is preserved on the high altar of the cathedral of Gerona; it is a casket, plated with silver gilt, and adorned with pearls, bearing an Arabic inscription invoking blessings upon the Prince of the Faithful, Hakam II., which reads rather curiously upon a Christian altar. The sword-hilts and jewels of the Moors were very elaborate, as the sword of Boabdil, the last King of Granada, shows. The Saracens were always renowned for their metal work, and even such small things as keys were beautifully ornamented. How exquisitely the Spanish Moors could chase bronze is proved by the engraving in chapter xi. of the beautiful mosque lamp which was made for Mohammed III. of Granada, and is still to be seen at Madrid. The delicacy of the open filigree work is only surpassed by similar work made at Damascus and Cairo. Over and over again we read the same Arabic inscription, the motto of the kings of Granada, "There is no conqueror but God." We have already spoken of the brass doors of the palaces of Cordova; and some remains of these are still to be seen in the Spanish cathedrals. Every one has heard of the Toledo sword-blades, and though the tempering of

HISPANO MORESCO LUSTRED PLATE, WITH ARMS OF LEON, CASTILE, AND ARAGON. (*In the South Kensington Museum.*)

steel is older in Spain than the invasion of the Arabs, the skill of the Toledo armourers was fostered by the Khalifs and Sultans of Cordova. Almeria, Seville, Murcia, and Granada were also famous places for armour and weapons. The will of Don Pedro in the fourteenth century runs: "I also endow my son with my Castilian sword, which I had made at Seville, ornamented with stones and gold." In arts, sciences, and civilization generally, the Moorish city of Cordova was indeed "the brightest splendour of the world."

IX.

THE PRIME MINISTER.

ABD-ER-RAHMĀN III. was the last great Sultan of Cordova, of the family of the Omeyyads. His son, Hakam II., was a bookworm, and although bookworms are very useful in their proper place, they seldom make great rulers. A king cannot be too highly educated; he may know everything under the sun, and, like several of the Cordovan Sultans, he may employ his leisure in music and poetry ; but he must not bury himself in his library, or care more for manuscripts than for campaigns, or prefer choice bookbinding to binding up the sore places of his subjects. Yet this was what Hakam did. He was not a weak man, or at all regardless of his great responsibilities ; but he was too much absorbed in his studies to care about the glories of war ; and his other delight, which consisted in building, was so far akin to his studious nature that it involved artistic tastes, which are often allied to those of literature. Hakam's peaceful, studious temperament did no great harm to the State. He was son enough of the Great Khalif to lead his armies against the Christians of Leon when they did not carry out their treaties ; and so overwhelming was the awe that his father had inspired, so universal the

ANCIENT KORAN CASE. (*Escurial Library.*)

sentiment of his crushing power, that the Christian princes of the north submitted to Hakam's interference with their affairs, and one of them even came to Cordova, and with many abject genuflexions implored the aid of the Sultan to restore him to his throne. Peace was soon signed between all the parties, and Hakam had leisure to collect his famous library. He sent agents to all parts of the East to buy rare manuscripts, and bring them back to Cordova. His representatives were constantly searching the booksellers' shops at Cairo and Damascus and Baghdad for rare volumes for the Sultan's library. When the book was not to be bought at any price, he would have it copied ; and sometimes he would even hear of a book which was only in the author's brain, and would send him a handsome present, and beg him to send the first copy to Cordova. By such means he gathered together no fewer than four hundred thousand books, and this at a time when printing was unknown, and every copy had to be painfully transcribed in the fine clear hand of the professional copyist. Not only did he possess all these volumes, but, unlike many collectors, he is said to have read them all, and even to have annotated them. So learned was he that his marginal notes were greatly prized by scholars of after times, and the destruction of a great part of his library by the Berbers was a serious loss to Arab literature.

It was possible for one successor of the Great Khalif to rest upon his father's laurels, and enjoy his studious tranquillity, while the enemy without was watching for an opportunity of renewing his attacks ;

but two such sovereigns would undo the great work which Abd-er-Rahmān had accomplished, and bring the Cordovan empire tumbling down to the ground again. Hakam II. only reigned fourteen years, and his son, Hishām II., was a boy of twelve when he ascended the throne. What the young Sultan might have been, had he been allowed fair play, no one can say; but it is recorded that he exhibited many signs of intelligence and sound judgment in his childhood, and showed some promise of following in the brilliant steps of his grandfather. Hakam's easy-going scholar's rule had, however, deprived his son and successor of any chance of real power. While the student Sultan was anxiously collating a manuscript, or giving directions to a copyist or bookbinder, the great officers of the State were gradually attaining a degree of authority which Abd-er-Rahmān III. would have instantly checked. The ladies of the Sultan's harīm also began to exercise an influence upon the government of the country. Abd-er-Rahmān built a city to please his wife, but he would have been very much astonished if Ez-Zahrā had ventured to dictate to him who was to be the prefect of police. When Hakam died, however, the harīm influence was very strong, and the Sultana Aurora, mother of the young Khalif Hishām, was perhaps the most important person in the State. There was one, however, a favourite of hers, who was destined soon to become even more influential. This was a young man called Ibn-Aby-Amir, or the "Son of the Father of Amir," but whom (since this is rather a roundabout name) we shall call by the title he afterwards adopted, when he had won many victories over the Christians—Almanzor, which means "the victorious by the grace of God." Almanzor started in life as an insignificant student at the university of Cordova, where his father was known as a learned lawyer of good but not influential family. The young man, however, had no intention of restricting his ambition to the modest elevation which his father had attained. While still a student he dreamed of power, and confidently predicted that one day he would be master of Andalusia; he even asked his schoolfellows—for they were little more than boys—what posts they would prefer to have when he came to power, and it is worth noticing that when that event came to pass he did not forget his promises. His career is an interesting example of what pluck, talent, and selfishness could do in a Moslem State, where the road to power was open to genius, however unpromising the beginnings. Almanzor, who was at first merely a professional letter-writer to the court servants, ingratiated himself with the Grand Chamberlain, who exercised the functions which would nowadays be held by a Prime Minister, and in due course he was appointed to some small offices about the court. Here his charm of manner and skilful flatteries gained him the favour of the ladies of the royal harīm, and especially of Aurora, who fell in love with the brilliant young man. Step by step, by dint of paying his court to the princesses, and making them magnificent presents (for which he had sometimes to draw upon public funds), he rose to higher offices; and by the age of thirty-one he enjoyed a comfortable plurality of posts, including that of superintendent of the property of the heir-apparent, a judgeship or two, and the office of commander of a division of the city guard. Everybody was charmed with his courtesy, his prodigal generosity, and the kindness with which he helped the unfortunate. He had already succeeded in attaching to himself a large number of persons, some of whom were of very high rank, when the death of the Khalif Hakam placed Aurora in a position of great importance, as mother of the boy Khalif, and gave Almanzor the opportunity he needed of making his power felt. The two worked together, and after establishing the child Hishām on the throne, which was only effected by the murder of a rival claimant, he quickly suppressed the conspiracy of the palace "Slavs," who would have nothing to say to the accession of Hishām. The head of the government was Mus-hafy, the chamberlain who had helped Almanzor to climb the first rung of the ladder of power; and his junior readily joined him in his policy. The repression of the Slavs, many of whom were now banished, made the two officials very popular with the people of Cordova, who cordially hated the foreign mercenaries. But this alliance was only for a time: as soon as he saw his way to get rid of the chamberlain, Almanzor was determined to do so without scruple. The first thing, however, was to increase his own popularity. An occasion immediately happened, which the young official boldly seized. The Christians were again becoming overweening on the northern marches, and the Chamberlain Mus-hafy, being no soldier, did not know how to cope with their aggressions. Almanzor, who had been a judge and an inspector, was no more a soldier than the chamberlain; but he came of a sound old stock, and his ancestor had been one of the few Arabs who had accompanied Tārik and his Berbers in the first invasion of Spain. Without a moment's hesitation or self-distrust, he volunteered to lead the army against the Christians; and so successful was the raid he made upon Leon, and so liberal was his *largesse* to the soldiery, that he returned to Cordova, not only triumphant—a civilian general—but also the idol of the army.

A second campaign was undertaken against the Christians of the north, in which the generalship was really done by Ghālib, the commander of the frontier forces, a brave officer, whom Almanzor adroitly made his friend. Ghālib protested so warmly that the victories were the fruit of the young civilian's talents, and vaunted his sagacity so highly, that the court and people came to believe that there lay a military genius under the cloak of the ex-lawyer—as, indeed, there was. Strengthened by this series of successes, and by Ghālib's support, Almanzor next ousted the son of the chamberlain from the post of prefect of Cordova, and took his place; and so admirably did he exert his authority, that never had the city been so orderly or the law so justly administered. Even his own son was beaten, till he died, because he had transgressed. His father, like Junius Brutus, allowed no exceptions in the execution of the law. By this policy he added to his laurels; he had already won over the army and pleased the populace, and now he had won the favour of all law-abiding citizens. The time had come for a great stroke of diplomacy. He played the chamberlain off against Ghālib so skilfully, that he widened the breach that already existed between the scarred man of arms and the nerveless clerk who held the functions of Prime Minister, and by inducing the former to throw over an engagement he was making with the chamberlain for an alliance between their families, and to give his daughter to Almanzor instead, he gave the last blow to the old minister. In 978, only two years after the death of Hakam, Almanzor had played his cards so ably, that he was in a position to accuse Mus-hafy of peculation—not without ample reason—and have him arrested, tried, and condemned. For five years the once powerful chamberlain led a wretched life at the heels of Almanzor, and then he died in prison, poisoned probably by his conqueror, in a state of utter destitution, covered only by an old tattered cloak of the jailor. Such was the fate of all who came between Almanzor and his ambition. The chamberlain, from the summit of glory and power, when thousands would come on bended knee to beg his favour, and when even an ex-king of Leon had sought humbly to kiss his hand, had been reduced to want and degradation by a young upstart whose insignificant origin had not crushed his genius.

That same day on which the chamberlain was disgraced, Almanzor stepped into his place. He was now at the height of power, and enjoyed the position of virtual ruler of all Mohammedan Spain. The government of Andalusia consisted of the Khalif in council; but Almanzor had buried the Khalif in his seraglio; and as for the Council of Vizirs who should advise him concerning affairs of State, Almanzor virtually united it in his own person. From his palace in the suburbs he ruled the whole kingdom; letters and proclamations were issued in his name; he was prayed for from the pulpits and commemorated on the coinage; and he even wore robes of gold tissue woven with his name, such as kings only were wont to wear. He was not, however, safe from the attacks of his enemies. Ambition brings its own dangers, and those who have been trampled upon are apt to turn and avenge themselves. Such was the case with Almanzor. One of the "Slavs," whom he had summarily deposed when they were planning a change in the succession, made an attempt to assassinate him; but it failed, and its author, along with a number of influential persons who had abetted the conspiracy, was arrested, condemned, and crucified.

In Cordova Almanzor was now supreme, for the young Khalif showed no symptoms of rebelling against the tutelage to which he was subjected, and the queen of the harīm, Aurora, was still the great minister's friend. One man only could pretend to any sort of equality with Almanzor, and this was Ghālib, his father-in-law. The army admired Almanzor, and wondered at his daring in taking the command of campaigns against the Christians without military experience; but they loved and adored Ghālib, as a type of the true warrior, bred to arms, and unconquerable in personal prowess. Ghālib was therefore a formidable rival, and Ghālib must be removed. The Prime Minister set about this task with his usual quiet determination. Whatever he undertook he carried out with the same immovable composure and iron will. A proof of his character was shown very strikingly one day, when he was seated with the Council of Vizirs, who formed the Cabinet of the Moorish government. They were discussing some public question, when a smell of burnt flesh rose in the chamber, and it was discovered that the minister's leg was being cauterized with red-hot iron while he was calmly debating the affairs of State! Such a man would find little difficulty in disposing of any obstacle—even General Ghālib. He laid his plans carefully, and they never failed. When his measures were a little too strong to be immediately approved by the people, he always had a plan ready for restoring the mob to acquiescence. Thus, when the revolt of several leading men had culminated in the attempted assassination already mentioned, he perceived that he had enemies among the theological and legal classes, and he lost no time in making his peace with them. Summoning a meeting of the chief doctrinal authorities, he asked them to make a list of those works on philosophy which they considered dangerous and heretical. The Moslems of Spain were famous for their rigid orthodoxy, and the philosophers received very harsh treatment from them. They soon decided upon what the Roman Catholic Church calls an "Index Expurgatorius," or list of condemned books, and Almanzor forthwith had the proscribed works publicly burnt. By this simple means, although really a man of broad views and perfectly tolerant of philosophical speculation, he succeeded in making himself the champion of orthodoxy: the theologians conspired no more against him.[1]

A man so fertile in expedients would not find much difficulty in getting rid of Ghālib. He first began a series of army reforms, by which he reduced the influence of individual commanders and gained for himself the devotion which had previously been bestowed upon captains of divisions. This he accomplished by drawing his recruits from Africa and from among the Christians of the north, who were of course without any prejudice in favour of any particular Moslem leader, and soon became attached to Almanzor, when they understood his liberality, and were convinced by repeated proofs of his military genius. He was a stern commander, and had been known to cut a man's head off with the culprit's own sword, because the same weapon had been seen gleaming in the dressed ranks when it should have been in its scabbard. But while a martinet in matters of drill and discipline, he was a father to his soldiers so long as they fought well and maintained order. His influence was unbounded. Once, when he sat in camp and saw his men in panic, running in, with the Christians at their heels, he threw himself from his throne, flung his helmet away, and sat down in the dust. The soldiers understood the despairing gesture of their general, and, suddenly turning about, fell upon the Christians, routed them, and pursued them even into the streets of Leon. Moreover, no one could lead them to such vast stores of booty as the man who made more than fifty successful campaigns against the princes of the

[1] Dozy: Hist. des Mus. d'Espagne, livre iii. ch. vi.-xii.

north. The army thus formed of new levies became devoted to their master, and Ghālib and his veterans of the frontier were speedily beaten; Ghālib himself died in an engagement. One other leader, Ja'far, the Prince of Zāb, threatened the peace of Almanzor by his extreme popularity with the troops; and he was presently invited to the minister's hall, made very drunk, and assassinated on his way home. This was by no means a solitary instance of Almanzor's treachery and bloodguiltiness; such acts deprive him of the title of hero to which his many brilliant qualities almost attain, and it is impossible to like him. Yet, with all his sternness and unscrupulousness, Almanzor brought Andalusia to a pitch of glory such as even the great Khalif, Abd-er-Rahmān III., had hardly contemplated. While keeping such hostile factions as remained in Cordova tranquil and powerless; whilst conciliating the people by making splendid additions to the great mosque of Cordova, when he found that they were beginning to grow indignant at the seclusion in which their young Khalif was kept, and were listening to the insinuations of Aurora and the palace party, who had grown tired or jealous of Almanzor; whilst overawing the Khalif himself by his personal influence; whilst keeping a watchful eye, that nothing escaped, upon every department of the administration, and devoting no little time to the cultivation of literature and poetry—amid all these various employments, this indefatigable man waged triumphant war in Africa and spread the dominion of the Khalif along the Barbary coast; and twice a year, in spring and autumn, led his troops, as a matter of course, against the Christians of Leon and Castile. Like a man of culture, he took his books along with his sword—his books were the poets who always accompanied his campaigns. Never was a general so constantly victorious. Supported by his hardy foreigners, and also by many Christians who were attracted by his pay and the sure prospect of booty, he carried fire and sword through the lands of the north. He captured Leon, and razed its massive walls and towers to the ground ; he seized Barcelona ; and, worst of all, he even ventured into the passes of Galicia, and levelled to the ground the splendid church of Santiago de Campostella, which was the focus of countless pilgrimages and almost formed the Kaaba of Europe. The shrine of St. James, however, where numerous miracles attested the presence of the saint's relics, was spared. It is said that when the conqueror entered the deserted city he found of all its inhabitants but a solitary monk, who still prayed before the holy shrine. "What doest thou here?" demanded Almanzor. "I am at my prayers," replied the old monk. His life was immediately spared, and a guard was set round the tomb to protect him and it from the violence of the soldiery, who proceeded to destroy everything else in the city. Almanzor well deserved his title of "Victorious," which was assumed after one of these campaigns. So long as his armies made their half-yearly expeditions, the Christian princes were paralysed, and Leon and the neighbouring country became a mere tributary province of the kingdom of Cordova. Castile, Barcelona, and Navarre were repeatedly defeated. He had taken the very capitals — Leon, Pamplona, Barcelona, and even Santiago de Campostella. Once he had brought the King of Navarre to his knees simply because the uncompromising Minister learned that there remained one captive Moslem woman in his kingdom. She was instantly delivered up, and many apologies were tendered for the inadvertence. Another time Almanzor found himself and his army cut off by the Christians, who had occupied an impregnable position in his rear, and barred his return to Cordova. Nothing daunted, he ordered his troops to foray the country round about, and collect materials for sheds, and implements of husbandry. Soon the Christians, who dared not attack, but believed they held the Moslems in their grasp, perceived them deliberately setting up barracks, and contentedly tilling the soil and preparing for the various operations of agriculture. Their astonished inquiries were answered by the cool reply, "We do not think it is worth while to go home, as the next campaign will begin almost immediately; so we are making ourselves comfortable for the interval!" Filled with consternation at the prospect of a permanent Moslem occupation, the Christians not only abandoned their strong position and allowed the enemy to go scot free, laden with booty, but even supplied them with baggage mules to carry off the spoils!

Almanzor, however, though invincible by man, was not proof against death. After a last victorious campaign against Castile, he was seized with mortal illness, and died at Medinaceli. The relief of the Christians is expressed in the simple comment of the monkish annalist: "In 1002 died Almanzor, and was buried in hell."

X.

THE BERBERS IN POWER.

THE best constituted countries will occasionally fall into anarchy when the will that has guided them is removed; and this is one of the strong arguments of those who hold that a State is best governed by the mass of its people. Keep a people in leading strings, it is said, and the moment the strings break, or are worn out, the people will not know where to go. The theory, however, is only a general statement of an obvious truth, and its application depends greatly upon the character of the people. Some nations seem always to need leading strings, and none has yet become absolutely independent of the guidance of a dominant mind; nor would such independence be desirable, unless a dead level of mediocrity be our ideal of a State. Andalusia, at all events, could not dispense with her leaders; and the instant her leader died, down fell the State. When "great Cæsar fell," then "I and you and all of us fell down," not so much for sympathy as incapacity. The multiplicity of mutually hostile parties and factions made anything resembling a settled constitution impossible in the dominion of the Moors. Only a strong hand could restrain the animosity of the opposing creeds and races in Andalusia; and those who have considered the character and history of Ireland, and the irreconcilable enmity which prevails between the north and the south in that island of factions, will allow that the Arabs were not the only people who found mixed races and religions impossible to gover with the smoothness of a homogeneous nation.

The history of Andalusia, so far as we have to it, has been a series of ups and downs. First we sa, a magnificent raid, led by born soldiers, ending in an unexpected conquest. Hardly was the peninsula won, when the jealousies and divisions of the various elements that made up the invading host bade fair to destroy the harvest just reaped by the sword. Then the strong man, the born king, appeared in the person of the first Abd-er-Rahmān, and Andalusia once more became, outwardly, one dominion. "O King, live for ever!" was the conventional form of address to the Persian monarch, and one is tempted to think that its realization might be the solution of all political troubles, provided the right king was chosen for immortality. The first king of Andalusia was naturally not immortal; and the consequence of his death was what always happens when a strong repressing force is withdrawn: the people fell again into civil war and anarchy. Yet again the God-gifted king came to rescue the nation. The Great Khalif imposed law and order throughout his dominions, beat back the invader, and trod the rebel under foot. For fifty years Andalusia was a paradise of peace and prosperity; had the third Abd-er-Rahmān been immortal she might have been peaceful to this day, and we should never have heard of the persecutions of Jews and Moors, of the terrible work of the Inquisition, or even (to come to very small things) the Carlists. It is a pity that such dreams cannot be true. But the Great Khalif had not left the country unprovided with a leader. A king had saved Spain twice, and now it was a prime minister who held the State together. Almanzor, the unconquerable minister, was able to make his masterful will felt to every corner of the peninsula ; but Almanzor, too, was mortal, and when he died, and (as the monk piously hoped) "was buried in hell," the land which owed him her prosperity and wealth, her perfect orderliness and security, became a prey to all the hostile forces which only his iron hand could repress. For eighty years Andalusia was torn to pieces by jealous chiefs, aggressive and quarrelsome tyrants, Moors, Arabs, Slavs, and Spaniards ; and though many of the old roots of dissension had been plucked up by time, and the jealousies that arose from memories of tribal glories were sometimes forgotten because men had lost their pedigrees, there were enough rivalries, personal, racial, and religious, to make Andalusia as much a hell upon earth as even the monkish chronicler could have desired for a burial-place for Almanzor.

For six years after the Prime Minister's death, his son Muzaffar maintained the unity of the kingdom. Then followed the deluge of greedy adventurers, rival khalifs, and impudent pretenders. The Spaniards, who formed after all the bulk of the population in which they were merged, loved to be ruled by a king ; they liked a dynasty, and were proud of the memories of the great Omeyyad house. The rule of a minister, however just and good, was not their idea of government; the king must rule by himself. So they rebelled against the authority of a second son of Almanzor, who had provoked them by publicly putting in his claim to succeed to the throne, and they insisted on the Khalif taking the reins of State into his own weak hands. The unfortunate Hishām, thus suddenly dragged out of the seclusion of his harīm, where he had been a happy prisoner for thirty years, in vain implored the people not to demand impossibilities of him; they would have him rule, and when it became clear to everybody that the feeble middle-aged man was as helpless as an infant, they made him abdicate, and set up another member of his family in his place. This was really the end of the Omeyyad dynasty of Andalusia. Khalif after khalif was set up for the next twenty years; one was the puppet of the Cordovans, another was the puppet of the Slav guards; a third was the puppet of the Berbers; a fourth was a sort of figure-head to mask the ambition of the ruler of Seville; but all were puppets of some faction, and had no vestige of real authority. The throne-room in the palace became the scene of murder after murder, as khalif succeeded khalif. One poor wretch hid himself in the oven of the bath-room, till he was discovered, dragged out, and butchered before the eyes of his successor, whose turn was not far off. Hishām II., the poor creature who had been kept in a state of perpetual infancy by Almanzor and the queen-mother Aurora, was forced to play his part in the raree-show. He was again set up, and again pulled down ; and the silken chains of his imprisonment among the beauties of his harīm were exchanged for the gloomy walls of a real dungeon. What became of him afterwards is unknown. His women said that he had contrived to escape, and had taken refuge in Asia, or at Mekka. The throne possessed few attractions for the miserable Khalif, who loved seclusion and pious duties ; and he must have known that his presence in Andalusia gave a rallying cry to ambitious partisans, and could only lead to further strife. It was natural that he should prefer to end his days in the exercise of devotion at the holy temple of Islam. An impostor, who closely resembled Hishām in person, set himself up as the Khalif at Seville, and was acknowledged as a convenient puppet by the powerful lord of that city ; but the real Hishām had disappeared for ever, and no one heard of him again.

How pitiful was the fate of the unhappy Omeyyads, who allowed the ferocious Moors, or Slavs, in turn, to use them as pieces on their chess-board, may be seen from what happened at the deposition of the third Hishām. By order of the chief men of the city, this mild and humane prince was dragged with his family to a dismal vault attached to the great mosque of Cordova. Here, in total darkness, half frozen with the cold and damp, and poisoned by the foul air of the place, the wretched Khalif sat, holding his only child, a little girl, to his breast, while his wives hung round him in scanty clothing, weeping, shivering, and dishevelled. They had been long without food, and their inhuman jailers had left them unnoticed for hours. The sheykhs then came to announce to Hishām the decision of the council which had been hastily summoned to debate upon his fate; but the poor Khalif, who was trying to restore a little warmth to the child in his arms, interrupted them: "Yes! yes! I will submit to their decision, whatever it is; but for God's sake get me some bread; this poor child is dying of hunger." The sheykhs were touched—they had not designed such torments—and the bread was brought. Then they began again: "Sire, they have determined that you shall be taken at daybreak to be imprisoned in such and such a fortress." "So be it," answered the Khalif; "I have only one favour to ask: permit us to have a lantern, for the darkness of this dismal place appals us." The lord spiritual and temporal of the Mussulmans of Spain had fallen to such straits that he had to beg for bread and a candle.

Such scenes as this were now frequent in Cordova. Each revolution brought its fresh crop of horrors. The people of Cordova, who had greatly increased in numbers, had also nourished those independent sentiments which the immense development of trade and manual industry, and the consequent creation of a prosperous artisan class, generally promote; and when they overturned Almanzor's dynasty, the mob broke out in the usual manner of mobs, and wreaked their vengeance by pillaging the beautiful palace which the great Minister had built in the neighbourhood of the capital for the use of himself and the government officials. When they had ransacked the priceless treasures of the palace, they abandoned it to

THE GIRALDA AT SEVILLE.

the flames. Massacres, plundering, and assassination went on unchecked for four days. Cordova became a shambles. Then the Berbers had their turn ; the imperious Slav guards, who had won the cordial detestation of the people, were succeeded by the brutal Berbers, who rioted in the plunder of the city. Wherever these barbarians went, slaughter, fire and outrage followed. Palace after palace was ransacked and burnt, and the lovely city of Ez-Zahrā, the delight of the Great Khalif, was captured by treachery, sacked, and set on fire, so that there remained of all the exquisite art that two khalifs had lavished upon its ornament nothing but a heap of blackened stones. Its garrison was put to the sword ; its inhabitants fled for refuge to the mosque ; but the Berbers had neither scruples nor bowels, and men, women, and children were butchered in the sacred precincts (1010).

While the capital was torn to pieces by savage bands of Slavs and Berbers, and was setting up one khalif after another, varying the family of Omeyya with that of Hammūd, or trying the effect of a governing town council, the provinces had long thrown off all allegiance to the central State. Every city or district had its own independent lord—so soon had the consolidating effects of Almanzor's rule disappeared. The Spaniards themselves enjoyed little of this sudden accession of small powers. They had to look on and lament, while foreigners divided their land among them. Berber generals fattened upon the South ; the Slavs subdued the East ; " the rest fell to parvenus or to the few noble families who had by some accident survived the blows which Abd-er-Rahmān III. and

Almanzor had dealt at the aristocracy. Cordova and Seville, the two most important cities of Andalus, had set up republics,"[1] in name, however, rather than fact; for the Moslem First Consul was a very close likeness of the Emperor. In the first half of the eleventh century some twenty independent dynasties came into power in as many towns or provinces, among which the Abbadites of Seville, the Hammūd family at Malaga and Algeciras, the Zirites at Granada, the Beny Hūd at Zaragoza, the Dhu-n-Nūn dynasty at Toledo, and the rulers of Valencia, Murcia, and Almeria, were the most important. Some of these dynasts were good rulers, most of them were sanguinary tyrants, but (curiously) not the less polished gentlemen, who delighted to do honour to learning and *belles lettres*, and made their courts the homes of poets and musicians. Mo'temid of Seville, for instance, was a prince of many accomplishments, yet he kept a garden of heads, cut off his enemies' shoulders, which he regarded with great pride and delight. As a whole, however, the country was a prey to disorder as intolerable and as dangerous as that which had prevailed when the Great Khalif came to the throne. It was not quite the same in character; for there was no great Christian rebellion like that of Ibn-Hafsūn; but the anarchy was as universal, and the danger of a total collapse more imminent than ever.

For the Christians of the north were now on the move. They saw their opportunity, and they made the most of it. Alfonso VI., who had united under

[1] Dozy, livre iii.

his sway the three kingdoms of the Asturias, Leon, and Castile, understood his part perfectly. He saw that he only had to allow the various Moslem princes rope enough, and they would proceed to hang themselves with the utmost expedition. These short-sighted tyrants, indeed, caring only for their petty individual power, and eagerly aiding in anything that could weaken their rivals, threw themselves at Alfonso's feet, and implored his assistance whenever they found themselves overmastered by a more powerful neighbour. Partly in consequence of acts of this kind, and partly in terror at the furious raids which the Castilians made throughout the country, even as far as the port of Cadiz, the Moslem States were almost all tributaries of the King of Castile, who took care to annually demand heavier and more heavy tribute, as the price of his friendship, in order to lay up stores for the great conquest which he had in mind. The north was poor, and with a fine irony he trusted to the immense contributions of his vassals among the Andalusian princes to provide the sinews of the war which should destroy them. Divided and jealous as were the Mohammedan dynasts, there was a limit to their patience. When Alfonso had bathed in the ocean by Hercules' Pillars, rejoicing that at last he had traversed all Spain and touched the watery border ; when he had established a garrison of more than twelve thousand daring men in the fortress of Aledo, in the very midst of the Moslem territories, whence they ruthlessly emerged to harry the whole country and commit every sort of savage outrage ; when Rodrigo Diaz de Bivar, " my Cid the

Challenger," had established himself in Valencia with his Castilians, and laid waste the neighbouring lands; when it became clear to every one that Alfonso meant nothing less than the reconquest of all Spain, and the extermination of all Moslems—then at last the Mohammedan princes awoke to their danger, and began to take measures for their defence. Helpless in themselves and, in spite of the common danger, despairing of any firm collected action among so many and such hostile factions, they took the only other course possible—they called in the aid of the foreigner. Some, indeed, foresaw dangers in such aid; but Mo'temid, the King of Seville, silenced them: "Better be a camel-driver in African deserts," he said, "than a swineherd in Castile!" The power they required was not far off. A new Berber revolution had taken place in North Africa, and a sect of fanatics, called the marabouts or saints (*Almoravides*, as the Spaniards named them), had conquered the whole country from Algiers to Senegal. They were much the same sort of people as Tārik and his followers, and they were ready enough to cross the water and conquer the fertile provinces of Spain. They made it a favour, indeed, and evinced supreme indifference to the attractions of Andalusia; but they came, and it was easy to see that they meant to stay.

When the Almoravides first came over like a cloud of locusts to devour the country thus offered to their appetite, they found the way perfectly open. The mass of the people of Andalusia rejoiced to see once more a strong arm coming to repress the disorder which had destroyed their well-being ever since the death of the great Almanzor; the petty tyrants either had invited them or could not resist them, and were, at all events, glad to see the Castilians successfully repelled. The Almoravide king, Yūsuf, the son of Teshfīn, after appropriating Algeciras, as a harbour and necessary basis of operations, marched unopposed through the provinces, and met Alfonso at Zallāka, or, as the Spaniards call it, Sacralias, near Badajoz, October 23, 1086. Alfonso, as he looked upon his own splendid army, exclaimed, "With men like these I would fight devils, angels, and ghosts!" Nevertheless he resorted to a ruse to score a surprise over the joint forces of the Berbers and Andalusians; but Yūsuf was not easily disconcerted. He took the Castilian army skilfully in front and rear, and, thus placed between two fires, in spite of the obstinate resistance which the tried warriors of Castile knew well how to offer, he crushed them utterly. Alfonso barely escaped with some five hundred horsemen. Many thousands of the best sword-arms in Castile lay stiff and nerveless on that fatal field.

After the victory, Yūsuf the Almoravide returned to Africa, leaving three thousand of his Berbers to help the Andalusians. He had promised to make no annexations, and, except in retaining the harbour of Algeciras, he had so far kept his word. The Andalusians were delighted with him; they praised his valour and exulted over the saving of the land; they admired his simple piety, which let him do nothing without the advice of his priests, and which had induced him to abolish all taxes in Spain except those few authorized by the Khalif Omar in the earliest days of Islam. The upper classes, indeed, ridiculed his ignorance and rough manners; he could speak but little Arabic, and when the poets recited their charming verses in his honour he generally missed the point of the compliment—no slight offence to the polished and elegant Andalusians, who never forgot their poetry even when they were up to their knees in blood. Yūsuf was to them a mere barbarian. But their contempt for his education did not greatly matter; they could not do without his sword, and the vast mass of the people, thinking rather of comfort than culture, were ready to receive him joyfully as sovereign of Andalusia. In 1090 the King of Seville again prayed the Almoravide to come over and help him against the Christians, who were as bold as ever, and carried on a perpetual guerilla warfare from their stronghold of Aledo. He acceded, with assumed unwillingness, and this time he directed his attacks quite as much against the Andalusian princes as against the Christians of Castile. These foolish tyrants dinned into his ears innumerable complaints against each other, and mutually betrayed themselves to such an extent, that Yūsuf very soon had grounds for distrusting the whole body of them. He had on his side the people, and, above all, the priests. These soon absolved him from his promise not to annex Andalusia, and even went so far as to urge him that it was his duty, in God's name, to restore peace and happiness to the distracted land. Always under the influence of his spiritual advisers, and sufficiently prompted by his own ambition without any such external impetus, Yūsuf readily fell in with this view, and before the year 1090 was out he had begun the subjugation of Spain. He entered Granada in November, and distributed its wonderful treasures—its diamonds, pearls, rubies, and other precious jewels, its splendid ornaments of gold and silver, its crystal cups, and gorgeous carpets, its unheard-of riches of every sort—among his officers, who had never in their lives seen anything approaching such magnificence. Tarīfa fell in December, and the next year saw the capture of Seville and many of the chief cities of Andalusia. An army sent by Alfonso, under the famous captain, Alvar Fañez, was defeated, and all the south lay at the feet of the Almoravides—save only Valencia, which no assault could carry so long as the Cid lived to direct the defence. In 1102, after the hero's death, Valencia succumbed, and now the whole of Mohammedan Spain, with the exception of Toledo, had become a province of the great African empire of the Almoravides.

The mass of the people had reason to be satisfied, for a time, with the result of their appeal to the foreigner. A minority, consisting of all the men of position and of education, were not so well pleased with the experiment. The reign of the Puritans had come, and without a Milton to soften its austerity. The poets and men of letters, who had thriven at the numerous little courts, where the most bloodthirsty despot had always a hearty and appreciative welcome for a man of genius, and would generally cap his verses with impromptu lines, were disgusted with the savage Berbers, who could not understand their refinements, and who, when they sometimes attempted to form themselves upon the model of the cultivated tyrants who had preceded them, made so poor an imitation that it was impossible to help laughing. The free-thinkers and men of broad views saw nothing very encouraging in the accession to power of the fanatical priests who formed the Almoravides' advisers, and who were not only rabidly opposed to anything that savoured of philosophy, but read their Koran exclusively through the spectacles of a single commentator. The Jews and Christians soon discovered what the tolerance of the Almoravides was : they were cruelly persecuted, massacred, or else transported. The old noble families, the few that remained, and the remnants of the petty princes, were in despair when they saw the stranger, whom they had bidden to their aid, taking up his permanent station in their dominions, and recalled with terror the doings of similar hordes of Berbers in the latter days of the Cordovan Khalifate. But the mass of the people were glad enough to see the Almoravides staying in the land ; their lives and goods were at last safe, which had never been the case when the country was cut up into a number of separate principalities, few of which were strong enough to protect their subjects outside the castle gates ; the roads were free from the brigands who had made travelling impossible for many years, and the Christians, instead of pouncing upon unsuspecting villages and harrying the land, were driven back to their own territory, where a wholesome dread of the Berbers, and a long strife among themselves, kept them at a safe distance. Order and tranquillity reigned for the moment ; the law was respected, and the people once more dreamed of wealth and happiness.

The dream was a delusion. There was no prosperity in store for the subjects of the Almoravides. What had happened to the Romans and the Goths now happened to the Berbers. They came to Spain hardy rough warriors, unused to ease or luxuries, delighting in feats of strength and prowess, filled with a fierce but simple zeal for their religion. They had not been long in the enjoyment of the fruits of their victory when all the demoralization which the soft luxuries of Capua brought upon the soldiers of Hannibal came also upon them. They lost their martial habits, their love of deeds of daring, their pleasure in enduring hardships in the brave way of war—they lost all their manliness with inconceivable rapidity. In twenty years there was no Berber army that could be trusted to repel the attacks of the Castilians; in its place was a disorganized crowd of sodden debauchees, miserable poltroons, who had drunk and fooled away their manhood's vigour and become slaves to all the appetites that make men cowards. Instead of preserving order, they had now become the disturbers of order; brigands, when they could pluck up courage to attack a peaceful traveller; thieves on all promising opportunities. The country was worse off than ever it had been, even under the petty tyrants. The enfeebled Berbers were at the beck and call of bad women and ambitious priests, and they would counterorder one day what they had commanded the day before. Such rulers do not rule for long. A great revolution was sapping the power of the Almoravides in Africa, and the Castilians under Alfonso the Battler resumed their raids into Andalusia. In 1125 they harried the south for a whole year. In 1133 they burnt the very suburbs of Cordova, Seville, and Carmona, and sacked Xeres and set it in a blaze. The Christian forays now extended from Leon to the Straits of Gibraltar, yet the besotted government did nothing to meet the danger. Exasperated at its feebleness, the people finally rose in their wrath and drove their impotent rulers from the land.

"At last," says the Arab historian, "when the people of Andalus saw that the empire of the Almoravides was falling to pieces, they waited no longer, but, casting away the mask of dissimulation, broke out into open rebellion. Every petty governor, chief, or man of influence, who could command a few followers and had a castle to retire to in case of need, styled himself Sultan, and assumed the other insignia of royalty; and Andalus had as many kings as there were towns in it. Ibn-Hamdīn rose at Cordova, Ibn-Maymūn at Cadiz, Ibn-Kāsy and Ibn-Wezīr Seddaray held the west, Lamtūny Granada, Ibn-Mardanīsh, Valencia; some Andalusians, others Berbers. All, however, shortly disappeared before the banners of Abd-el-Mumin, who deprived every one of them of their dominions, and subjected the whole of Andalus to his rule." Abd-el-Mumin was the leader of the Almohades, who succeeded to the Almoravide power in Africa and Spain.

XI.

MY CID THE CHALLENGER.

IT is time to glance at the opponents of the Moors in the North. We have seen how Pelayo gathered together the remnant of the Goths in the inaccessible caves and fastnesses of the Asturian mountains; how this remnant soon advanced beyond its early boundaries, and, taking courage from the indifference or the disunion of the Berber tribes who were quartered on the frontiers of the Mohammedan dominions, gradually recovered most of the territory north of the Sierra de Guadarrama, and there established the kingdom of Leon and the county of Castile; while the separate kingdom of Navarre arose further east, beneath the Pyrenees. We have also seen how these Christian kingdoms were in a state of almost constant war with their Moorish neighbours, and might have been seriously dangerous but for the no less constant divisions which neutralized the various Christian States. So long as the kingdom of Cordova remained strong and undivided, while the Christians of Leon, Castile, and Navarre wasted their vigour in civil wars, the Moors were fully equal to the task of preserving their dominions. But when the kingdom of Cordova fell, and Andalusia became a prey to petty dynasties, each of which thought first of its own interests, and then perhaps of the interests of the Mohammedan power at large, the Christians became more venturesome, and were enabled to wring from the Moors a considerable accession of territory. During the confusion of the eleventh century, when almost every city in Andalusia formed a State by itself, we have seen that the Christians scoured the land of the Moslems with their victorious armies, and exacted tribute from many of the most important Moorish princes. At this time Fernando the First had united the greater part of the north under his own sceptre. He had combined the conflicting provinces of Leon and Castile, and incorporated the Asturias and Galicia in his dominions. Fernando was undoubtedly the most powerful monarch in all Spain at this time; he had annexed Lormego, Viseu, and Coimbra in Portugal, and took tribute from the kings of Zaragoza, Toledo, Badajoz, and Seville; and though his imprudent division of his dominions among his three sons and two daughters involved the north in a series of civil wars after his death, Alfonso VI. "the Valiant" eventually succeeded in cementing the scattered fragments together again, and henceforward the progress of the Christian power in Spain was inevitable. It was only the immense bribes of the Mohammedan princes (who paid blackmail to a fabulous amount to buy off the Christians), and the armies of the Almoravides in the background, that prevented the entire reconquest of Andalusia by the Christians at this period of Moorish weakness. As it was, the Moors were in no sense their own masters;

BOTICA DE LOS TEMPLARIOS, TOLEDO.

they were harassed between the dread of Alfonso and the scarcely less alarming supremacy of their Almoravide ally ; and in the end they had to succumb to the latter. At this time we find the Christians interfering in most of the political affairs of the Mohammedan states ; Christian armies overrunning their territories and demanding heavy tribute for their goodwill ; and so complicated became the alliances between the two parties that many Christian mercenaries were to be found in the armies of the Moors, vigorously assisting in campaigns of devastation and sacrilege through Christian provinces, while Moors were ready to join the Castilians against their fellow-Moslems. It was, in short, a time of adventurers, of paid mercenaries, of men who fought for personal interest and profit, instead of for king and country.

We should make a great mistake if we regarded the warriors of Leon and Castile as anything approaching an ideal of knightly honour and chivalry, and a still greater error would be to imagine them polished, cultivated gentlemen. The Christians of the north formed the most striking possible contrast to their Moorish rivals. The Arabs, rough tribesmen as they had been at their first arrival, had softened, by contact with the Andalusians and by their own natural disposition to enjoyment and luxury, into a highly civilized people, delighting in poetry and elegant literature, devoted to the pursuit of learning, and, above all, determined to enjoy life to the utmost. Their intellectual tastes were unusually fine and delicate ; they were moved by emotions which could only be felt by men of taste and *savoir vivre.* They were romantic, imaginative, poetical, speculative, and would bestow on a well-turned epigram what would have sufficed to pay a regiment of soldiers. The most tyrannical and bloodthirsty among their despots was held in some contempt if he were not also something of a poet, or at least instinctively appreciative of polished wit and courtly eloquence. Music, oratory, as well as the severer pursuits of science, seemed to come naturally to this brilliant people; and they possessed in a high degree that quality of critical perception and delicate appreciation of the finer shades of expression which in the present day we associate with the French nation.

The Christians of the north were as unlike this as can well be conceived. Though descended from an older kingdom, the northern states had most of the qualities of new nations. They were rude and uncultivated; few of their princes possessed the elements of what could be called education, and they were too poor to indulge in the refined luxuries of the Moorish sovereigns. The Christians were simply rough warriors, as fond of fighting as even their Moslem antagonists, but even better prepared by their hard and necessarily self-denying lives for the endurance of long campaigns and the performance of desperate deeds of valour. They had no idea of the high standard of chivalrous conduct which poets afterwards infused into their histories; they were men of the sword, and little besides. Their poverty made them any man's servants; they sold their valour to him who paid them best; they fought to get a livelihood. We have seen how the great minister Almanzor won his victories against Leon and took Santiago with the aid of a large contingent of the Leonese themselves, who perceived clearly enough on which side their fortunes were to be made. The history of the eleventh century in Spain is full of such examples of the employment of Christian *chevaliers d'industrie* by Moorish princes; but of these none has ever attained such celebrity as the Cid, the national hero of Spain.

The Cid's proper name was Rodrigo Diaz of Bivar, and he was called the *Cid* because that was the title which his Moorish followers naturally gave him. A Mohammedan gentleman is still addressed in Egypt and elsewhere by the title *Sīd*, which is a corruption of the word *Seyyid*, meaning "master." The Cid, or "master," was also styled *Campeador*, which signifies "champion," or, more accurately, "challenger," because his exceeding prowess made him the natural challenger in those single combats which in Spanish wars commonly preceded a general engagement between two armies. A famous warrior would advance before the ranks, as Goliath of Gath stood forth before the armies of Israel, and challenge the opposing forces to send him out a champion; and none was more renowned for his triumphs in this manner of warfare than Rodrigo Diaz, "myo Cid el Campeador," as the old chronicler affectionately calls him. It is not easy to decide how much of the splendid history which has gathered round the exploits of the Cid is true. The Christian chroniclers stopped at nothing when they began to describe their national hero; and the enthusiasm that did not shrink from relating how the King of Leon seized Paris, and conquered the French, Germans, Italians, and even the Persians, can be trusted still less when it sounds the glories of the beloved Cid. The Spanish ballads surround their hero with a saintly aureole of all the virtues, and forget that many of these virtues would not have been understood or appreciated by the Cid himself or his contemporaries in Castile. The Arabic writers are generally more trustworthy, but their judgment can hardly have been unbiassed when they spoke of a Christian who worked such misery to the Moslems of Valencia as did the famous Campeador. Yet even they call him a "miracle of God."

In this critical age we are frequently obliged to abandon with regret the most charming traditions of our childhood's histories; and the Cid has not been spared. A special book has been written by an eminent Orientalist to prove that the redoubtable Challenger was by no means the hero he was supposed to be; that he was treacherous and cruel, a violator of altars, and a breaker of his own good faith. Professor Dozy maintains that the romantic history of the Cid is a tissue of inventions, and he has written an account of "the real Cid" to counteract these misleading narratives. He founds his criticisms mainly on the Arabic historians, in whom, despite their national and religious bias, he places as blind a reliance as less learned people have placed in the *Chronicle of the Cid.* Yet it is surprising how trifling are the differences that can be detected between his "real Cid" and that romantic *Chronicle of the Cid,* the substance of which was compiled by Alfonso the Learned only half a century after the Cid's death, and which Robert Southey translated into English in 1805 with such skill and charm of style that his version has ever since been almost as much a classic as the original. Every one can separate for himself the obviously legendary incidents in the delightful old *Chronicle* without any assistance from the Arabic historians, who deal chiefly with one period alone of the Cid's career; and the best popular account of the hero, in discriminating hands and with due allowances, is still Southey's fascinating *Chronicle*. The Cid of the *Chronicle* is not at all the same as the Cid of the Romances; and while we cheerfully abandon the latter immaculate personage, we may still believe in the former. Of course our Cid had his faults, and was guilty of not a few thoroughly indefensible acts. He was no very orthodox champion of the faith, for he fought as well for the Moors as for the Christians, and would as dispassionately rob a church as a mosque. But all this is clear enough to any one who reads the *Chronicle*, and it does not make the Cid anything but what he always was—a hero of the rude days of yore. If we are to limit our definition of heroism to characters that display all Christian virtues, long-suffering, gentleness, and pity, we shall have to dismiss most of our old friends. Achilles was not very gentle or compassionate when he dragged the body of Hector round the walls of Troy: but Achilles is the hero of the Iliad. Nine out of ten of the heroes of antiquity committed a host of acts which we moderns, with our superfine sensibilities, call cruel, ungenerous, even dastardly. It is a pure perversion of history to apply latter-day codes of morality to the heroes of bygone ages. Let us admit that they are not all gold; and then let us delight in their great deeds, the mighty swing of their sword-arm, the crushing shock of their onset, their tall stature and flashing eyes as they ride to meet their foes. We do not expect them to be philosophers or strict advocates of the theories of political economy. We are quite satisfied with them as they are: heroes, —brave, gallant leaders of men.

The Cid was a real hero to the Spaniards: first, because he fought so magnificently, and that used once to be title enough to reverence; secondly, because, like the mythical Bernardo del Carpio and the real Fernando Gonzalez, he was the champion of Castile, and had bearded the King of Leon, and thus represented the immemorial jealousy which the Castilians entertained for the powerful neighbours who absorbed their province; and thirdly, because the minstrels forgot his long alliance with the Moors, or contrived to give it a disinterested aspect, and remembered him only as the great champion of the Christian people against the infidels. But the very cause which specially commended him to the Castilians, his insubordination to King Alfonso, made him a less perfect hero to the writer of the *Cronica General*, from which the *Chronicle of the Cid* was extracted. That writer or compiler, Alfonso the Learned, King of Leon and Castile, could not approve the haughty independence of the Cid towards his own forerunner the sixth Alfonso. Hence in

Southey's version of the *Chronicle* (which is enriched with many extracts from the *Poem of the Cid* and other sources) we have a check upon the excessive adulation of the ballads and romances. There is no lack of details in the work which are anything but creditable to the Cid; but, nevertheless, the true heroic character, with all its faults and limitations, is well sustained, and the record forms a wonderfully interesting picture of a stirring time and the greatest figure among the Spanish chevaliers.

The story of the Cid would fill a volume by itself; all we can attempt here is to extract a few of the most striking passages of the *Chronicle*. The youth of the hero is, to a large extent, merged in myth; he first comes into historical documents in 1064, when, though scarcely more than twenty, he had already won his title of Challenger by a triumphant single combat with a knight of Navarre, and was soon afterwards appointed commander-in-chief of the forces of Castile. He helped Sancho of Castile to overcome his brother Alfonso of Leon, by a surprise which savoured strongly of treachery, but which passed for good strategy in those rough-and-ready times. After the murder of Sancho by Bellido, under the walls of Zamora, the Cid passed into the service of his successor, the very Alfonso whom he had before driven into exile. The king at first welcomed the invincible knight of Castile to his court, and married him to his own cousin; but jealous rivals poisoned his mind, already filled with the memory of past wrongs, against Rodrigo (or Ruy Diez, as he is styled in the *Chronicle*), and in 1081 the Cid was banished from his dominions. The *Chronicle* must tell the story of his farewells:

"And the Cid sent for all his friends and his kinsmen and vassals, and told them how King Don Alfonso had banished him from the land, and asked of them who would follow him into banishment, and who would remain at home. Then Alvar Fañez, who was his cousin-german, came forward and said, Cid, we will all go with you, through desert and through peopled country, and never fail you. In your service will we spend our mules and horses, our wealth and our garments, and ever while we live be unto you loyal friends and vassals. And they all confirmed what Alvar Fañez had said ; and the Cid thanked them for their love, and said that there might come a time in which he should guerdon them.

"And as he was about to depart he looked back upon his own home, and when he saw his hall deserted the household chests unfastened the doors open, no cloaks hanging up, no seats in the porch, no hawks upon the perches, the tears came into his eyes, and he said, My enemies have done this. . . . God be praised for all things. And he turned toward the East and knelt and said, Holy Mary Mother, and all Saints, pray to God for me, that He may give me strength to destroy all the Pagans, and to win enough from them to requite my friends therewith, and all those who follow and help me. Then he called for Alvar Fañez and said unto him, Cousin, the poor have no part in the wrong which the king hath done us ; see now that no wrong be done unto them along our road ; and he called for his horse. And then an old woman who was standing at her door said, Go in a lucky minute, and make spoil of whatever you wish.

And with this proverb he rode on, saying, Friends, by God's good pleasure we shall return to Castile with great honour and great gain. And as they went out from Bivar they had a crow on their right hand, and when they came to Burgos they had a crow on the left.

"My Cid Ruydiez entered Burgos, having sixty streamers in his company. And men and women went forth to see him, and the men of Burgos and the women of Burgos were at their windows, weeping, so great was their sorrow; and they said with one accord, *Dios!* how good a vassal if he had but a good lord! and willingly would each have bade him come in, but no one dared so to do. For King Don Alfonso in his anger had sent letters to Burgos, saying that no man should give the Cid a lodging; and that whosoever disobeyed should lose all that he had, and moreover the eyes in his head. Great sorrow had these Christian folk at this, and they hid themselves when he came near them because they did not dare speak to him; and my Cid went to his Posada, and when he came to the door he found it fastened for fear of the king. And his people called out with a loud voice, but they within made no answer. And the Cid rode up to the door, and took his foot out of the stirrup, and gave it a kick, but the door did not open with it, for it was well secured; a little girl of nine years old then came out of one of the houses and said unto him, O Cid, the king hath forbidden us to receive you. We dare not open our doors to you, for we should lose our houses and all that we have, and the eyes in our head. Cid, our evil would not help you, but God and all His saints be with you. And when she had said this she returned into the house. And when the Cid knew what the king had done he turned away from the door and rode up to St. Mary's, and there he alighted and knelt down, and prayed with all his heart; and then he mounted again and rode out of the town, and pitched his tent near Arlanzon, upon the Glera, that is to say, upon the sands. My Cid Ruydiez, he who in a happy hour first girt on his sword, took up his lodging upon the sands, because there was none who would receive him within his door. He had a good company round about him, and there he lodged as if he had been among the mountains. . . .

"The cocks were crowing amain, and the day began to break, when the good Campeador reached St. Pedro's. The Abbot Don Sisebuto was saying matins, and Doña Ximena (the Cid's wife) and five of her ladies of good lineage were with him, praying to God and St. Peter to help my Cid. And when he called at the gate and they knew his voice, *Dios!* what a joyful man was the Abbot Don Sisebuto! Out into the courtyard they went with torches and with tapers, and the Abbot gave thanks to God that he now beheld the face of my Cid. And the Cid told him all that had befallen him, and how he was a banished man; and he gave him fifty marks for himself, and a hundred for Doña Ximena and her children. Abbot, said he, I leave two little girls behind me, whom I commend to your care. Take you care of them and of my wife and of her ladies: when this money be gone, if it be not enough, supply them abundantly; for every mark which you expend upon them I will give the monastery four. And the Abbot promised to do this with a right good will. Then Doña Ximena came up, and her daughters with her, each of them borne in arms, and she knelt down on both her knees before her husband, weeping bitterly, and she would have kissed his hand; and she said to him, Lo, now you are banished from the land by mischief-making men, and here am I with your daughters, who are little ones and of tender years, and we and you must be parted, even in your life-time. For the love of St. Mary tell me now what we shall do. And the Cid took the children in his arms, and held them to his heart and wept, for he dearly loved them. Please God and St. Mary, said he, I shall yet live to give these my daughters in marriage with my own hands, and to do you service yet, my honoured wife, whom I have ever loved even as my own soul.

"A great feast did they make that day in the monastery for the good Campeador, and the bells of St. Pedro's rung merrily. Meantime the tidings had gone through Castile how my Cid was banished from the land, and great was the sorrow of the people. Some left their houses to follow him, others forsook their honourable offices which they held. And that day a hundred and fifteen knights assembled at the bridge of Arlanzon, all in quest of my Cid; and there Martin Antolinez joined them, and they rode on together to St. Pedro's. And when he of Bivar knew what a goodly company were coming to join him, he rejoiced in his own strength, and rode out to meet them and greeted them full courteously; and they kissed his hand, and he said to them, I pray to God that I may one day requite ye well, because ye have forsaken your houses and your heritages for my sake, and I trust that I shall pay ye twofold. Six days of the term allotted were now gone, and three only remained: if after that time he should be found within the king's dominions, neither for gold nor for silver could he then escape. That day they feasted together, and when it was evening the Cid distributed among them all that he had, giving to each man according to what he was; and he told them that they must meet at mass after matins, and depart at that early hour. Before the cock crew they were ready, and the Abbot said the mass of the Holy Trinity, and when it was done they left the church and went to horse. And my Cid embraced Doña Ximena and his daughters, and blessed them; and the parting between them was like separating the nail from the quick flesh: and he wept and continued to look round after them. Then Alvar Fañez came up to him and said, Where is your courage, my Cid? In a good hour were you born of woman. Think of our road now; these sorrows will yet be turned into joy."

The Cid offered his services to the Moorish King of Zaragoza, the most powerful of the northern Moslem princes; and they were joyfully accepted. At the head of his own followers, who were the more devoted to him since they lived by the booty he procured them, he made a raid through Aragon, and so rapid was his riding that he harried a vast tract of country in five days, and was off before the Christians could sound the alarm. He led the Moors against the

Count of Barcelona, won a signal victory, and made the Count his ally. How the Cid and his merry men triumphed in the battle-field, let the *Chronicle* again relate:

"Pero Bermudez could not bear this, but holding the banner in his hand, he cried, God help you, Cid Campeador; I shall put your banner in the middle of that main body; and you who are bound to stand by it—I shall see how you will succour it. And he began to prick forward. And the Campeador called unto him to stop as he loved him, but Pero Bermudez replied he would stop for nothing, and away he spurred and carried his banner into the middle of the great body of the Moors. And the Moors fell upon him that they might win the banner, and beset him on all sides, giving him many and great blows to beat him down; nevertheless, his arms were proof, and they could not pierce them, neither could they beat him down, nor force the banner from him, for he was a right brave man and a strong and a good horseman, and of great heart. And when the Cid saw him thus beset, he called to his people to move on and help him. Then placed they their shields before their hearts, and lowered their lances with the streamers thereon, and, bending forward, rode on. Three hundred lances were they, each with its pendant, and every man at the first charge slew his Moor. Smite them, knights, for the love of charity! cried the Campeador. I am Ruydiez, the Cid of Bivar! Many a shield was pierced that day, and many a false corselet was broken, and many a white streamer dyed with blood, and many a horse left without a rider.

The misbelievers called on Mahomet, and the Christians on Santiago, and the noise of the tambours and of the trumpets was so great that none could hear his neighbour. And my Cid and his company succoured Pero Bermudez, and they rode through the host of the Moors, slaying as they went, and they rode back again in like manner ; thirteen hundred did they kill in this guise. If you would know who they were, who were the good men of that day, it behoves me to tell you, for though they are departed, it is not fitting that the names of those who have done well should die, nor would they who have done well themselves, or who hope so to do, think it right ; for good men would not be so bound to do well if their good feats should be kept silent. There was my Cid, the good man in battle, who fought well upon his gilt saddle ; and Alvar Fañez Minaya, and Martin Antolinez the Burgalese of prowess, and Muno Gustios, and Martin Munoz who held Montemayor, and Alvar Alvarez, and Alvar Salvadores, and Galin Garcia the good one of Aragon, and Felez Munoz the nephew of the Campeador. Wherever my Cid went, the Moors made a path before him, for he smote them down without mercy. And while the battle still continued, the Moors killed the horse of Alvar Fañez, and his lance was broken, and he fought bravely with his sword afoot. And my Cid, seeing him, came up to an Alguazil, who rode upon a good horse, and smote him with his sword under the right arm, so that he cut him through and through, and he gave the horse to Alvar Fañez, saying, Mount Minaya, for you are my right hand."

GATE OF SERRANO, VALENCIA.

call cruel, ungenerous, even dastardly. It is a pure perversion of history to apply latter-day codes of morality to the heroes of bygone ages. Let us admit that they are not all gold; and then let us delight in their great deeds, the mighty swing of their sword-arm, the crushing shock of their onset, their tall stature and flashing eyes as they ride to meet their foes. We do not expect them to be philosophers or strict advocates of the theories of political economy. We are quite satisfied with them as they are: heroes, —brave, gallant leaders of men.

The Cid was a real hero to the Spaniards: first, because he fought so magnificently, and that used once to be title enough to reverence; secondly, because, like the mythical Bernardo del Carpio and the real Fernando Gonzalez, he was the champion of Castile, and had bearded the King of Leon, and thus represented the immemorial jealousy which the Castilians entertained for the powerful neighbours who absorbed their province; and thirdly, because the minstrels forgot his long alliance with the Moors, or contrived to give it a disinterested aspect, and remembered him only as the great champion of the Christian people against the infidels. But the very cause which specially commended him to the Castilians, his insubordination to King Alfonso, made him a less perfect hero to the writer of the *Cronica General*, from which the *Chronicle of the Cid* was extracted. That writer or compiler, Alfonso the Learned, King of Leon and Castile, could not approve the haughty independence of the Cid towards his own forerunner the sixth Alfonso. Hence in

Southey's version of the *Chronicle* (which is enriched with many extracts from the *Poem of the Cid* and other sources) we have a check upon the excessive adulation of the ballads and romances. There is no lack of details in the work which are anything but creditable to the Cid; but, nevertheless, the true heroic character, with all its faults and limitations, is well sustained, and the record forms a wonderfully interesting picture of a stirring time and the greatest figure among the Spanish chevaliers.

The story of the Cid would fill a volume by itself; all we can attempt here is to extract a few of the most striking passages of the *Chronicle*. The youth of the hero is, to a large extent, merged in myth; he first comes into historical documents in 1064, when, though scarcely more than twenty, he had already won his title of Challenger by a triumphant single combat with a knight of Navarre, and was soon afterwards appointed commander-in-chief of the forces of Castile. He helped Sancho of Castile to overcome his brother Alfonso of Leon, by a surprise which savoured strongly of treachery, but which passed for good strategy in those rough-and-ready times. After the murder of Sancho by Bellido, under the walls of Zamora, the Cid passed into the service of his successor, the very Alfonso whom he had before driven into exile. The king at first welcomed the invincible knight of Castile to his court, and married him to his own cousin; but jealous rivals poisoned his mind, already filled with the memory of past wrongs, against Rodrigo (or Ruy Diez, as he is styled in the *Chronicle*), and in 1081 the Cid was banished from his dominions. The *Chronicle* must tell the story of his farewells:

decreed that all the Cid's conquests should be his own property. In other words, he recognized the Cid as an almost independent prince. Almost immediately, however, the king became again suspicious of his powerful vassal, and seized the opportunity of the Cid's absence in the north to besiege his peculiar possession, the city of Valencia. When the Campeador heard this he was very wroth, and, by way of retaliation, carried fire and sword through Alfonso's districts of Najera and Calahorra, razed Logroño to the ground, and, in the words of the old Latin *Gesta*, "with terrible and impious despoilment he wasted and harried the land, and stripped it bare of its riches and seized them for himself." Alfonso hastily abandoned the siege of Valencia, and returned to defend his own country. But the Cid, having effected his purpose, came back another way, and found the gates of Valencia closed against him.

Then began that memorable siege of nine months, during which the people of Valencia suffered agonies of hunger and thirst, while the Cid maintained his remorseless leaguer round the walls. The besieged were reduced to the agonies of starvation, and those who rushed out, or were thrust forth as useless burdens by the townspeople, were massacred or sold into slavery by the Cid's soldiers. It is even said by the Moorish historians that the Cid had many of them burnt alive. The *Chronicle* pathetically records: "Now there was no food to be bought in the city, and the people were in the waves of death; and men were seen to drop and die in the streets." Thus wrote a poet of the devoted city:

"Valencia! Valencia! trouble is come upon thee, and thou art in the hour of death; and if peradventure thou shouldst escape, it will be a wonder to all that shall behold thee.

"But if ever God hath shown mercy to any place, let Him be pleased to show mercy unto thee; for thy name was joy, and all Moors delighted in thee and took their pleasure in thee.

"And if it should please God utterly to destroy thee now, it will be for thy great sins, and for the great presumption which thou hadst in thy pride.

"The four corner stones whereon thou art founded would meet together and lament for thee, if they could!

"Thy strong wall which is founded upon these four stones trembles, and is about to fall, and hath lost all its strength.

"Thy lofty and fair towers which were seen from far, and rejoiced the hearts of the people, . . . little by little they are falling.

"Thy white battlements which glittered afar off, have lost their truth with which they shone like the sunbeams.

"Thy noble river Guadalaviar, with all the other waters with which thou hast been served so well, have left their channel, and now they run where they should not.

"Thy water-courses, which were so clear and of such great profit to so many, for lack of cleansing are choked with mud.

"Thy pleasant gardens which were round about thee; . . . the ravenous wolf hath gnawn at the roots, and the trees can yield thee no fruit.

"Thy goodly fields, with so many and such fair flowers, wherein thy people were wont to take their pastime, are all dried up.

"Thy noble harbour, which was so great honour to thee, is deprived of all the nobleness which was wont to come into it for thy sake.

"The fire hath laid waste the lands of which thou wert called Mistress, and the great smoke thereof reacheth thee.

"There is no medicine for thy sore infirmity, and the physicians despair of healing thee.

"Valencia! Valencia! from a broken heart have I uttered all these things which I have said of thee.

"And this grief would I keep unto myself that none should know it, if it were not needful that it should be known to all."

At last, in June, 1094, Valencia surrendered, and the Cid stood once more upon her towers and ramparts. He made hard conditions with the people, many of whom he sent away to the suburbs to make room for his Castilians. But if he was harsh and not quite honest in his dealings with the vanquished, his triumph was stained by no wholesale butchery. The people were sometimes ruined; but their lives, except their leader's, were safe. The Cid had now attained the summit of his power. He sent for his wife and daughters from the abbey, and established himself permanently as King of Valencia and suzerain of the country round about. The King of Aragon besought his alliance. He exacted heavy tribute from his neighbours; his revenue included 120,000 pieces of gold yearly from Valencia, 10,000 from the lord of

Albarracin, 10,000 from the heir of Alpuente, 6,000 from the Master of Murviedro, and so forth. He dreamed of reconquering all Andalusia. "One Roderick," he said, "lost Spain; another shall recover it." When the Almoravides came against him, he put them to rout. The *Chronicle* tells the story:

"Day is gone, and night is come. At cock-crow they all assembled together in the Church of St. Pedro, and the Bishop Don Hieronymo sang mass, and they were shriven and assoyled and howselled. Great was the absolution which the bishop gave them: He who shall die, said he, fighting face forward, I will take his sins, and God shall have his soul. Then said he, A boon, Cid Don Rodrigo; I have sung mass to you this morning: let me have the giving the first wounds in this battle! and the Cid granted him this boon in the name of God. Then, being all ready, they went out through the gate which is called the Gate of the Snake, for the greatest power of the Moors was on that side, leaving good men to guard the gates. Alvar Fañez and his company were already gone forth, and had laid their ambush. Four thousand, lacking thirty, were they who went out with my Cid, with a good will, to attack fifty thousand. They went through all the narrow places and bad passes, and, leaving the ambush on the left, struck to the right hand, so as to get the Moors between them and the town. And the Cid put his battles in good array, and bade Pero Bermudez bear his banner. When the Moors saw this they were greatly amazed; and they harnessed themselves in great haste, and came out of their tents. Then the Cid bade his banner move on, and the Bishop Don Hieronymo pricked forward with his company, and laid on with such guise, that the hosts were soon mingled together. Then might you have seen many a horse running about the field with the saddle under his belly, and many a horseman in evil plight upon the ground. Great was the smiting and slaying in short time; but by reason that the Moors were so great a number, they bore hard upon the Christians, and were in the hour of overcoming them. And the Cid began to encourage them with a loud voice, shouting God and Santiago! And Alvar Fañez at this time issued out from ambush, and fell upon them, on the side which was nearest the sea; and the Moors thought that a great power had arrived to the Cid's succour, and they were dismayed, and began to fly. And the Cid and his people pursued, punishing them in a bad way. If we should wish to tell you how every one behaved himself in this battle, it is a thing which could not be done, for all did so well that no man can relate their feats. And the Cid Ruydiez did so well, and made such mortality among the Moors, that the blood ran from his wrist to his elbow! Great pleasure had he in his horse Bavieca that day, to find himself so well mounted. And in the pursuit he came up to King Yusuf, and smote him three times; but the king escaped from under the sword, for the horse of the Cid passed on in his course, and when he turned, the king being on a fleet horse, was far off, so that he might not be overtaken; and he got into a castle called Guyera, for so far did the Christians pursue them, smiting and slaying, and giving them no respite, so that hardly fifteen thousand escaped of fifty that they were."

TOMB OF THE CID AT SAN PEDRO DE CARDEÑA.

But the fortune of war is fickle. The troops of the Cid were defeated at last by the invaders; and the Campeador died of grief in July, 1099. They took his body and embalmed it, and kept vigil by its side; then, in the legend of the poets, they did as the Cid had bidden them: they set him upon his good horse Bavieca, and fastened the saddle well, so that he sat erect, with his countenance unchanged, his eyes bright and fair, and his beard flowing down his breast, and his trusty sword Tizona in his hand. No one would have known that he was dead. And they led Bavieca out of the city: Pero Bermudez in front with the banner of the Cid and five hundred knights to guard it, and Doña Ximena behind with her company and escort. Slowly they cut a path through the besiegers, and took the road to Castile, leaving the Moors in sore amazement at their strange departure: for they did not know that the Cid was dead. But the body of the hero was set in an ivory chair beside the great altar of San Pedro de Cardeña, under a canopy whereon were blazoned the arms of Castile and Leon, Navarre and Aragon, and of the Cid Campeador. Ten years the Cid sat upright beside the altar, his face still noble and comely, when the signs of death at last began to appear; so they buried him before the altar, where Doña Ximena already lay; and they left him in the vault, still upright in the ivory chair, still in his princely robes with the sword Tizona in his hand,—still the great Campeador whose dinted shield and banner of victory hung desolate over his tomb.

XII.

THE KINGDOM OF GRANADA.

WITH such soldiers as the Cid, and such kings as Fernando and Alfonso, the recovery of all Spain by the Christians was only a matter of time. Every nation, it appears, has its time of growth and its period of efflorescence, after which comes the age of decay. As Greece fell, as Rome fell, as every ancient kingdom the world has known has risen, triumphed, and fallen, so fell the Moors in Spain. Their time was now near at hand. They had been divided and undisciplined before the Almoravide annexation: they were not less so when their Berber masters had been expelled. But hardly had the Almoravides disappeared, when a new enemy came on the scene. The Almohades, or fanatical "Unitarians," who had overthrown the power of the Almoravides in Africa, resolved to imitate their vanquished predecessors by including Andalusia in their empire. The dissensions among the princes of the long-shattered kingdom of the Moors made the task an easy one. In 1145 the Almohades took Algeciras; in 1146 they occupied Seville and Malaga, and the next four years saw Cordova and the rest of southern Spain united under their sway. Some princes, indeed, held out for a while,

BANNER OF THE ALMOHADES.

but the hordes of African fanatics were too overpowering for any single chief to make a protracted stand against them.

The Almohades, however, had no thought of making Andalusia the centre of their government. They ruled it from Africa, and the consequence was that their hold upon Spain was weak. The disturbed provinces of Andalusia were not easily to be retained by princes who contented themselves with deputies sent from Morocco, and with an occasional expedition to repel the attacks of the Christians. When they came in force their efforts were generally crowned with success. They won a splendid victory over the Christians in 1195 at Alarcos, near Badajoz, where thousands of the enemy were slain, and immense spoils fell into the hands of the fanatics. But the fortune of war changed when, in 1212, the disastrous field of Las Navas decided the fate of the Almohades. Of 600,000 men, few escaped to tell the tale of slaughter. City after city fell into the hands of the Christians; and family dissensions among the foreigners, and the attacks of rival dynasties in Africa, enabled the chiefs of Andalusia, who had grown impatient of the spasmodic rule of their foreign masters, in 1235, to drive the Almohades out of the peninsula. An Arab chief, Ibn-Hūd, then made himself master of most of the south of Spain, and even of Ceuta in Africa; but he died in 1238, and the command of Andalusia now devolved upon the Beny-Nasr of Granada.

The kingdom of Granada was the last bulwark of the Moors in Spain. It was not much that was now left to them. Between 1238 and 1260, Fernando III. of Castile and Jayme I. of Aragon conquered Valencia, Cordova, Seville, and Murcia; and the rule of the Moors was now restricted to the present province of Granada, *i.e.*, the country about the Sierra Nevada and the sea coast from Almeria to Gibraltar Within this limit, however, their kingdom was destined to endure for another two centuries and a half Though hemmed in on all sides, the Moors were well served by soldiers. The people of the conquered cities, the most valiant warriors of the vanquished Moslem states, came to place their swords at the disposal of the one remaining Mohammedan king. Fifty thousand Moors are recorded to have fled to his protection from Valencia, and three hundred thousand from Seville, Xeres, and Cadiz. Nevertheless, Granada was forced to become tributary to the Castilian crown. The founder of the dynasty of the Beny-Nasr, an Arab named Ibn-el-Ahmar, or the "Red man," because of his fair skin and hair, was a vigorous sovereign, but he could not withstand the power of the Christians, who now held nearly the whole of Spain. He paid homage and tribute to Fernando and his son Alfonso the Learned, not, however, without more than one struggle to free himself from their yoke; and from that time forward Granada with its surrounding territory was generally let alone by the Christian kings, who had enough to do to settle their already vast acquired territory and to do away with local pretenders. From time to time the Moors made war upon their Christian neighbours, but eventually they had to make up their minds to a secondary position.

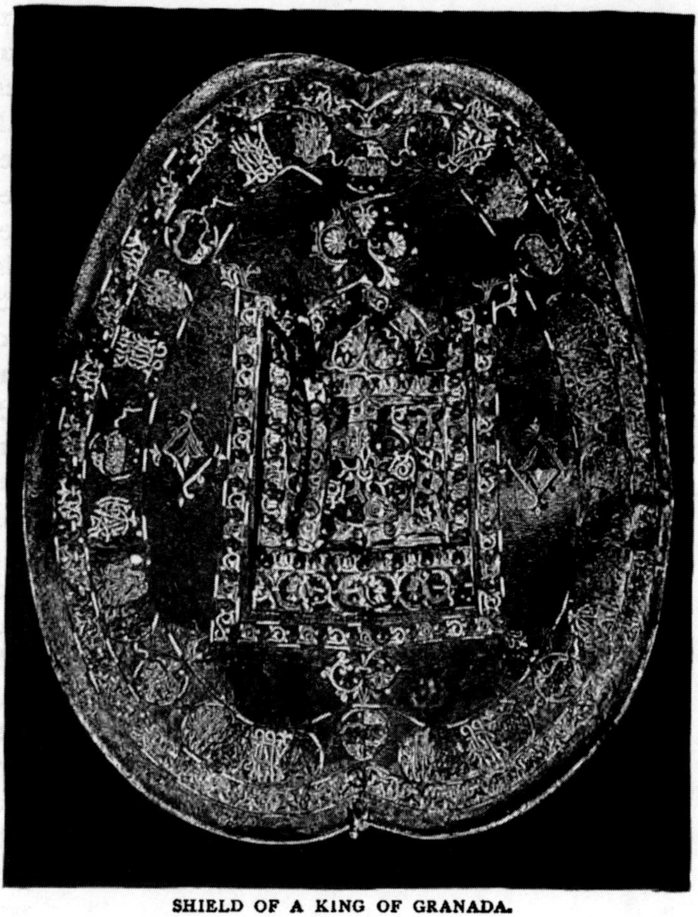

SHIELD OF A KING OF GRANADA.

The sum of twelve thousand gold ducats was the tribute paid by Mohammed X., in 1463, as a condition of peace. During these two centuries the Moorish territory had suffered little diminution. Gibraltar had been lost and won and lost again; other places, notably Algeciras, had become part of the Christian dominions; but the general extent of the Moslem realm remained in the third quarter of the fifteenth century much what it had been in the first half of the thirteenth.

During this period of comparative tranquillity, Granada had taken the place of Cordova as the home of the arts and sciences. Its architects were renowned throughout Europe; they had built the marvellous "Red Palace," *Alhambra*, so called from the colour of the ferruginous soil on which it stands, and they had covered it with the splendid gold ornament and Arabesque mouldings which are still the wonder of artists of all countries.[1] Granada itself, with its two castles, was a pearl of price. It stands on the border of a rich plain, the famous "Vega," lying at the feet of the snowy "mountains of the moon," the Sierra Nevada. From the heights of the city, and still better from the Alhambra, which stands sentinel over the plain like the Acropolis of Athens, the eye ranges over this beautiful Vega, with its streams and vineyards, its orchards and orange groves. No city in Andalusia was more favoured in site or climate; the breezes

[1] The Alhambra was begun in the thirteenth century and completed in the fourteenth. Washington Irving, who visited it in 1829, in company with Prince Dolgorouki, has given an interesting account of his life there, which combines the romance and the history of the place.

from the snow mountains made the hottest summer tolerable, and the land was fertile beyond compare.

The site chosen by the Moors for their celebrated Red Palace is a terrace bounded by precipitous ravines, at the foot of which, to the north, flow the waters of the river Darro. Solid walls of stone covered with stucco, and strengthened at frequent intervals by towers, surround the terrace. The enclosed space is somewhat of the form of a lanceolate leaf lying on the table-land, with its greatest length (about half a mile) from east to west.

The visitor finds his way into the enclosure through a massive embattled tower of orange and red pierced by the Gate of Justice under which the khalifs, like the judges of the Hebrews, were wont to sit in judgment. Twenty-eight feet above the pavement, over the horseshoe arch, a cabalistic key and a gigantic hand are carved on two stones. Once inside the walls, the visitor finds himself in a square, on one side of which is an unfinished palace designed by Charles the Fifth. The corridor through which entrance is now gained to the Alhambra crosses an angle of this ruined structure and admits the visitor to the Court of the Myrtles, so called from the profusion of those shrubs which adorn its sides. A narrow passage ushers us into a court one hundred and forty feet long, and half as broad, flooded with sunlight and gay with gold-fish, which disport themselves in a long pond that fills the larger part of the space. Pillars and galleries adorn the sides and ends of the enclosure, and on the north the great square tower of Comares rises against the horizon. The

THE COURT OF THE LIONS IN THE ALHAMBRA.

court is a place of peace; the water scarcely makes a ripple as it gently oozes into the ample reservoir, and leaves it without a gurgle; the multitudinous gold-fishes gleam and glitter in the profusion of sunshine; no suggestion of the outer world penetrates the stillness.

All is calm, but it is not the dull, cold calm of ruin and death; we can but feel a sense of companionship with the former masters of the palace and the grounds. We walk through the Barca, or boat-shaped ante-chamber, to the Hall of the Ambassadors, and imagine the khalif of the Omeyyads seated upon his throne at the end; while we gaze up into the lofty dome and allow our eyes to wander about the great apartment as we admire the medallions, the graceful characters of the Arabic inscriptions, the delicate patterns of the plaster-work with which the walls are adorned; the balconies, the white, blue, and gold of the cornice and ceiling; the circles, crowns, and stars moulded to imitate the vault of heaven. We stop before the window looking over the Darro to think how Ayesha once let Boabdil down in a basket from it five centuries ago; how Charles the Fifth said of the unfortunate Moor, "Ill-fated was the man who lost all this!" We bring up before us the discoverer of America, as tradition paints him, pleading in this place with the good Isabella for gracious permission to add another jewel to her crown—the bright gem of a New World. We climb to the terraced roof of the tower, following the narrow windings of the steep stairway once trodden by fair lady and gallant prince as they hastened to the lofty battlement to watch the approach of some army or anxiously to study the progress of a battle on the Vega. Our eyes search the broad expanse for that bridge of Pinos where Moor and Christian more than once fought for the mastery. We remember that it was at that spot that Isabella's messenger overtook the despairing Columbus, as he conveyed to him the queen's recall, when the mariner was plodding towards other realms to carry his bold proposition to other and, as he hoped, more gracious sovereigns. We care not that it is tradition only which fixes the spot; tradition and romance are a portion of the charm of the Alhambra.

In our search through the intricate plan of the pile, we find ourselves in the boudoir of the Sultana, the windows of which command the same prospect over the Vega, a scene to which distance lends its greatest charm. We are reminded on every side of the luxury of the olden time, when we see the apertures in the white marble floor near the entrance, through which perfumes arose from drugs, which tradition says were burned beneath the floor to make the air of the lady's apartment redolent with their sweet scents. We look down into the garden of the Lindaraja, upon which Irving also looked when he occupied those apartments which have become historic on his account. The garden itself is scarcely worthy of notice, for it is a little-cultivated court; but near by are the baths of the Sultans, with their delicate filigree work, intricate tracery, and brilliant mosaics. There is the fountain which ripples in gentle cadence, as if keeping time to the harmony that the musicians poured down from the balconies when the ladies of the harim enjoyed the pleasures of the Oriental bath, or rested themselves upon cloth of gold. Each bath, cut from a single mass of white marble, was placed in its own vaulted chamber, and lighted through open-work of stars and roses.

Perhaps the most celebrated portion of the entire palace is the Court of the Lions, which occupies a space somewhat smaller than that of the Court of the Myrtles. One hundred and twenty-eight white marble columns, arranged by threes and fours in symmetrical fashion, support galleries which rise to no very lofty height; but the extreme gracefulness and elegance of their varied capitals, the delicate traceries, the remnants of gold and colour, the raised orange-shaped cupolas, the graceful minarets, the innumerable arches, beautiful in their labyrinthine design, the empty basin into which the twelve stiff and unnatural "lions" once poured their constant streams of cooling waters, the alabaster reservoir, constitute a whole that poetry and romance have lauded even to extravagance.

From this beautiful court, through a door ornamented with rare designs, one is ushered into the Hall of the Abencerrages, named from the legend that in its precincts the chiefs of that family were beheaded by order of Boabdil. Convenient spots in the stone floor are exhibited to credulous visitors as evidences that the blood was there spilt. The sweet and peaceful light which enters the apartment by sixteen airy windows in the star-shaped stalactite roof, illuminating its arches ornamented in azure and scarlet, seem all inappropriate to such a scene of slaughter, and charity would lead us, if history did not, to doubt that the stain should rest upon the memory of Boabdil.

Time would fail us to go through all the courts and halls of the comprehensive building, and we must make our way over the road that crosses the ravine of Los Molinos, bordered with figs and pistachios, laurels and roses, to the other palace, the Generalife, or "Garden of the Surveyor." This is the "Country House" of the greater palace, and, so far as the exterior of the building is concerned, presents the usual simplicity of Oriental structures. Here are the walls without windows, the terraces, the galleries, the arcades—all of which are in a state of ruin. The delicate arabesques are covered with thick layers of whitewash; the fine sculptures have disappeared, and the internal beauty of the edifice has long since departed; but the charm of the gardens and waters remains. A rapid stream runs through an artificial channel of marble the entire length of the enclosure under a series of arcades and leafy screens formed by curiously twisted yews, while cypresses and orange trees cast their cooling shadows upon the waters. Jets and fountains, rapid-flowing streams and placid ponds, little basins nestling under ancient bays, murmuring rivulets and winding courses reflecting the blue of the sky, are intermingled with the utmost perfection of skill, and the combination forms one of the most charming effects that can be imagined.

"Here," says Irving, "is everything to delight a southern voluptuary: fruits, flowers, fragrance, green arbours and myrtle hedges, delicate air and gushing water. Here I had an opportunity of witnessing

GARDEN OF THE GENERALIFE, GRANADA.

those scenes which painters are fond of depicting about southern palaces and gardens. It was the saint's day of the Count's daughter, and she had brought up several of her youthful companions from Granada to sport away a long summer's day among the breezy halls and bowers of the Moorish palace. A visit to the Generalife was the morning's entertainment. Here some of the gay companions dispersed themselves in groups about the green walks, the bright fountains, the flight of Italian steps, the noble terraces, and marble balustrades. Others, among whom I was one, took their seats in an open gallery or colonnade, commanding a vast prospect; with the Alhambra, the city, and the Vega far below, and the distant horizon of mountains—a dreamy world, all glimmering to the eye in summer sunshine. While thus seated, the all-pervading tinkling of the guitar and click on the castanets came stealing up the valley of the Darro, and half-way down the mountain we descried a festive party under the trees enjoying themselves in true Andalusian style; some lying on the grass, others dancing to the music."

From the ruined building one gains, perhaps, the most satisfactory view of the Alhambra, as its reddish line of half-demolished walls is traced along the undulations of the mountain on which it stands; while the white ridges of the Sierra Nevada furnish a magnificent background for the picture, and set off the heavy mass of the unfinished palace of Charles the Fifth.

Two centuries of prosperity, with a powerful Christian State almost at bow-shot, were as much as the

Moors had any right to expect; and towards the third quarter of the fifteenth century there were signs that their knell was about to sound. The union of Aragon with Castile by the marriage of Ferdinand and Isabella was the note of doom. Two such sovereigns could not long leave the Moors undisturbed in their corner of the peninsula. Muley Aly, generally known by his surname, Abu-l-Hasan (which the Spaniards change into Alboacen, and many English writers into Aben Hasan), who was of a fiery and warlike nature, resolved to be beforehand with their Catholic majesties in opening the game of war. He refused to pay the customary tribute, and when the ambassador of Ferdinand came to insist, he made answer: "Tell your sovereigns that the kings of Granada who paid tribute are dead: our mint now coins nothing but sword-blades!" To make his meaning unmistakable, he proceeded to carry a raid into the lands of the Christians. Zahara was the spot he selected for attack. A gifted American author has told the story of the last wars of the Moors in his own eloquent style; and we must follow Washington Irving in relating the assault of Zahara.[1]

"In the year of our Lord one thousand four hundred and eighty one, and but a night or two after the festival of the most blessed Nativity, the inhabitants of Zahara were sunk in profound sleep; the very sentinel had deserted his post, and sought shelter from a tempest which had raged without for three nights in succession; for it appeared but little probable that an enemy would be abroad during such an uproar of

[1] Chronicle of the Conquest of Granada, chap. iv.

the elements. But evil spirits work best during a storm. In the midst of the night an uproar rose within the walls of Zahara, more awful than the raging of the storm. A fearful alarm-cry, 'The Moor!' 'The Moor!' resounded through the streets, mingled with the clash of arms, the shriek of anguish, and the shout of victory. Muley Abu-l-Hasan, at the head of a powerful force, had hurried from Granada, and passed unobserved through the mountains in the obscurity of the tempest. While the storm pelted the sentinel from his post and howled around tower and battlement, the Moors had planted their scaling-ladders, and mounted securely into both town and castle. The garrison was unsuspicious of danger until battle and massacre burst forth within its very walls. It seemed to the affrighted inhabitants as if the fiends of the air had come upon the wings of the wind, and possessed themselves of tower and turret. The war-cry resounded on every side, shout answering shout, above, below, on the battlements of the castle, in the streets of the town; the foe was in all parts, wrapped in obscurity, but acting in concert by the aid of preconcerted signals. Starting from sleep, the soldiers were intercepted and cut down as they rushed from their quarters; or, if they escaped, they knew not where to assemble, or where to strike. Whenever lights appeared, the flashing scimitar was at its deadly work, and all who attempted resistance fell beneath its edge. In a little while the struggle was at an end. Those who were not slain took refuge in the secret places of their houses, or gave themselves up as captives. The clash of arms ceased, and the storm continued its howling, mingled with the occasional shout of the Moorish soldiery roaming in search of plunder. While the inhabitants were trembling for their fate, a trumpet resounded through the streets, summoning them all to assemble, unarmed, in the public square. Here they were surrounded by soldiery, and strictly guarded until daybreak. When the day dawned, it was piteous to behold this once prosperous community, which had lain down to rest in peaceful security, now crowded together without distinction of age, or rank, or sex, and almost without raiment, during the severity of a winter storm. The fierce Muley Abu-l-Hasan turned a deaf ear to all remonstrances, and ordered them to be conducted captives to Granada. Leaving a strong garrison in both town and castle, with orders to put them in a complete state of defence, he returned flushed with victory to his capital, entering it at the head of his troops, laden with spoil, and bearing in triumph the banners and pennons taken at Zahara. While preparations were making for jousts and other festivities in honour of this victory over the Christians, the captives of Zahara arrived—a wretched train of men, women, and children, worn out with fatigue and haggard with despair, and driven like cattle into the city gates by a detachment of Moorish soldiery."

The civilized people of Granada were shocked at the cruelty of Abu-l-Hasan, and felt that this was the beginning of the end. "Woe to Granada!" they cried. "The hour of its desolation is at hand. The ruins of Zahara will fall upon our own heads!"

Retribution was not far off. The redoubtable

Marquess of Cadiz captured the castle of Alhama by surprise, and thus planted a Christian garrison in the heart of the Moslem territory, within a short distance of Granada itself. In vain did Muley Abu-l-Hasan invest the captured castle; the Christians within performed prodigies of valour in its defence, and held the place till their friends came to their support. *Ay de mi Alhama!* "Woe for my Alhama!" was the cry that arose in Granada; "Alhama is fallen; the key of Granada is in the hands of the infidels!" Byron has made every one familiar with the plaintive ballad which he mistranslated:

> Pasavase el rey Moro
> Por la ciudad de Granada,
> Desde las puertas de Elvira
> Hasta las de Bivarambla.
> Ay de mi Alhama!

Henceforward, the castle proved a sore thorn in the side of the Moorish kings; for thence the brave Count of Tendillo harried the Vega and wrought infinite destruction. "It was a pleasing and refreshing sight," says the Jesuit chronicler[1] invented by Washington Irving, "to behold the pious knight and his followers returning from one of these crusades, leaving the rich land of the infidel in smoking desolation behind them: to behold the long line of mules and asses laden with the plunder of the

[1] Mr. Irving says of his "chronicler": "In constructing my chronicle, I adopted the fiction of a Spanish monk as a chronicler. Fray Antonio Agapida was intended as a personification of the monkish zealots who hovered about the sovereigns in their campaigns, marring the chivalry of the camp by the bigotry of the cloister, and chronicling in rapturous strains every act of intolerance towards the Moors." (Introduction to the revised edition of the Conquest of Granada, 1850.)

Gentiles, the hosts of captive Moors, men, women, and children; droves of sturdy beeves, lowing kine and bleating sheep—all winding up the steep acclivity to the gates of Alhama, pricked on by the Catholic soldiery. . . . It was an awful spectacle at night to behold the volumes of black smoke, mingled with lurid flames, that rose from the burning suburbs, and the women on the walls of the towns wringing their hands and shrieking at the desolation of their dwellings."

Inflamed by their respective conquests, both sides busied themselves in raids such as these, with little result, save general devastation and exasperation. The Christians at last attempted a movement on a larger scale. They resolved to invade the province of Malaga, and, marshalling the forces of the south, led by the Marquess of Cadiz and other noted warriors, they set out upon their fateful march. "It was on a Wednesday[a] that the pranking army of high-mettled warriors issued forth from the ancient gates of Antequera. They marched all day and night, making their way secretly, as they supposed, through the passes of the mountains. As the tract of country they intended to maraud was far in the Moorish territories, near the coast of the Mediterranean, they did not arrive there till late in the following day. In passing through these stern and lofty mountains, their path was often along the bottom of a barranca, or deep rocky valley, with a scanty stream dashing along it, among the loose rocks and stones which it had broken and rolled

[a] Washington Irving: Conquest of Granada, chap. xii.

down in the time of its autumnal violence. Sometimes their road was a mere rambla, or dry bed of a torrent cut deep into the mountains and filled with their shattered fragments. These barrancas and ramblas were overhung by immense cliffs and precipices, forming the lurking-places of ambuscades during the wars between the Moors and Spaniards, as in after times they have become the favourite haunts of robbers to waylay the unfortunate traveller.

"As the sun went down, the cavaliers came to a lofty part of the mountains, commanding, to their right, a distant glimpse of a part of the fair Vega of Malaga, with the blue Mediterranean beyond, and they hailed it with exultation, as a glimpse of the promised land. As the night closed in they reached the chain of little valleys and hamlets, locked up among those rocky heights, and known among the Moors by the name of Axarquia. Here their vaunting hopes were destined to meet the first disappointment. The inhabitants had heard of their approach; they had conveyed away their cattle and effects, and with their wives and children had taken refuge in the towers and fortresses of the mountains.

"Enraged at their disappointment, the troops set fire to the deserted houses, and pressed forward, hoping for better fortune as they advanced. Don Alonzo de Aguilar, and the other cavaliers in the van-guard, spread out their forces to lay waste the country, capturing a few lingering herds of cattle, with the Moorish peasants who were driving them to some place of safety.

"While this marauding party carried fire and sword

in the advance, and lit up the mountain cliffs with the flames of the hamlets, the Master of Santiago, who brought up the rear-guard, maintained strict order, keeping his knights together in martial array, ready for attack or defence should an enemy appear. The men-at-arms of the Holy Brotherhood attempted to roam in quest of booty; but he called them back and rebuked them severely.

"At last they came to a part of the mountain completely broken up by barrancas and ramblas, of vast depth, and shagged with rocks and precipices. It was impossible to maintain the order of march; the horses had no room for action, and were scarcely manageable, having to scramble from rock to rock, and up and down frightful declivities, where there was scarce footing for a mountain goat. Passing by a burning village, the light of tne flames revealed their perplexed situation. The Moors, who had taken refuge in a watch-tower on an impending height, shouted with exultation when they looked down upon these glistening cavaliers, struggling and stumbling among the rocks. Sallying forth from their tower, they took possession of the cliffs which overhung the ravine, and hurled darts and stones upon the enemy.

"In this extremity the Master of Santiago despatched messengers in search of succour. The Marquess of Cadiz, like a loyal companion-in-arms, hastened to his aid with his cavalry. His approach checked the assaults of the enemy, and the master was at length enabled to extricate his troops from the defile. . . .

"The Adalides, or guides, were ordered to lead the way out of this place of carnage. These, thinking to conduct them by the most secure route, led them by a steep and rocky pass, difficult for the foot soldiers, but almost impracticable to the cavalry. It was overhung with precipices, from whence showers of stones and arrows were poured upon them, accompanied by savage yells, which appalled the stoutest heart. In some places they could pass but one at a time, and were often transpierced, horse and rider, by the Moorish darts, impeding the progress of their comrades by their dying struggles. The surrounding precipices were lit up by a thousand alarm fires; every crag and cliff had its flames, by the light of which they beheld their foes bounding from rock to rock, and looking more like fiends than mortal men. Either through terror and confusion, or through real ignorance of the country, their guides, instead of conducting them out of the mountains, led them deeper into their fatal recesses. The morning dawned upon them in a narrow rambla; its bottom formed of broken rocks, where once had raved along the mountain torrent; while above them beetled huge arid cliffs, over the brows of which they beheld the turbaned heads of their fierce and exulting foes. . . .

"All day they made ineffectual attempts to extricate themselves from the mountains. Columns of smoke rose from the heights where, in the preceding night, had blazed the alarm fire. The mountaineers assembled from every direction: they swarmed at every pass, getting in the advance of the Christians, and garrisoning the cliffs, like so many towers and battlements.

"Night closed again upon the Christians, when they

were shut up in a narrow valley traversed by a deep stream, and surrounded by precipices which seemed to reach the sky, and on which the alarm fires blazed and flared. Suddenly a new cry was heard resounding along the valley. Ez-Zagel! Ez-Zagel! echoed from cliff to cliff. 'What cry is that?' said the master of Santiago. 'It is the war-cry of Ez-Zagel, the Moorish general,' said an old Castilian soldier; 'he must be coming in person with the troops of Malaga.'

"The worthy Master turned to his knights: 'Let us die,' said he, 'making a road with our hearts, since we cannot with our swords. Let us scale the mountains, and sell our lives dearly, instead of staying here to be tamely butchered.'

"So saying, he turned his steed against the mountain, and spurred him up its flinty side. Horse and foot followed his example, eager, if they could not escape, to have at least a dying blow at the enemy. As they struggled up the height, a tremendous storm of darts and stones was showered upon them by the Moors. Sometimes a fragment of rock came bounding and thundering down, ploughing its way through the centre of their host. The foot soldiers, faint with weariness and hunger, or crippled by wounds, held by the tails and manes of their horses, to aid them in their ascent, while the horses, losing their footing among the loose stones, or receiving some sudden wound, tumbled down the steep declivity, steed, rider, and soldier rolling from crag to crag, until they were dashed to pieces in the valley. In this desperate struggle the Alferez, or standard-bearer of the Master, with his standard was lost, as were many of his relations and dearest friends. At length he succeeded in attaining the crest of the mountain; but it was only to be plunged in new difficulties. A wilderness of rocks and rugged dells lay before him, beset by cruel foes. Having neither banner nor trumpet, by which to rally his troops, they wandered apart, each intent upon saving himself from the precipices of the mountains and the darts of the enemy. When the pious Master of Santiago beheld the scattered fragments of his late gallant force he could not restrain his grief. 'O God!' exclaimed he, 'great is Thy anger this day against Thy servants! Thou hast converted the cowardice of these infidels into desperate valour, and hast made peasants and boors victorious over armed men of battle!'

"He would fain have kept his foot soldiers and gathered them together, and have made head against the enemy; but those around him entreated him to think only of his personal safety. To remain was to perish without striking a blow; to escape was to preserve a life that might be devoted to vengeance on the Moors. The Master reluctantly yielded to their advice. 'O Lord of Hosts,' exclaimed he again, 'from Thy wrath do I fly, not from these infidels. They are but instruments in Thy hands to chastise us for our sins!' So saying, he sent the guides in advance, and, putting spurs to his horse, dashed through a defile of the mountain before the Moors could intercept him. The moment the Master put his horse to speed, his troops scattered in all directions: some endeavoured to follow his traces, but were confounded among the intricacies of the mountain.

They fled hither and thither, many perishing among the precipices, others being slain by the Moors, and others taken prisoners."

The horrors of that night among the mountains of Malaga were not speedily forgotten by the Christians. They burned for vengeance; and when "Boabdil" (properly Abu-Abdallah), the King of Granada, who had temporarily ousted his father from the sovereignty, sallied forth on a sweeping raid into the lands of the Christians, they took a signal revenge. Boabdil marched secretly by night; but his movements were not long undetected. Beacon fires blazed from the hill-tops, and the Count of Cabra, aroused by their flames, sounded the alarm, and assembled the chiefs of the district. They fell upon the Moors near Lucena, and, aided by the cover of the woods, made so skilful an attack, that the enemy turned. "Remember the mountains of Malaga!" was the ominous cry, as the Christian knights set spurs to their horses in pursuit of the Moslems: with shouts of St. James they dashed upon them, and the retreat became an utter rout. When the fugitives entered the gates of Granada a great wave of lamentation passed through the city: "Beautiful Granada, how is thy glory faded! The flower of thy chivalry lies low in the land of the stranger; no longer does the Bivarambla echo to the tramp of steed and sound of trumpet; no longer is it crowded with thy youthful nobles, gloriously arrayed for the tilt and tourney. Beautiful Granada! the soft note of the lute no longer floats through thy moonlit streets; the serenade is no more heard beneath thy balconies; the lively castanet is silent upon thy

A WINDOW IN THE ALHAMBRA.

hills; the graceful dance of the Zambra is no more seen beneath thy bowers. Beautiful Granada! why is the Alhambra so forlorn and desolate? The orange and myrtle still breathe their perfumes into its silken chambers; the nightingale still sings within its groves; its marble halls are still refreshed with the plash of fountains and the gush of limpid rills! Alas! the countenance of the king no longer shines within those halls. The light of the Alhambra is set for ever!"

Boabdil, indeed, had been made prisoner and was now a captive on his way to Cordova, while Ferdinand ravaged the Vega, and old Muley Abu-l-Hasan, who now returned to his kingdom, ground his teeth in impotent rage behind his stout ramparts.

XIII.

THE FALL OF GRANADA.

THE capture of Boabdil by the Christian sovereigns was a fatal blow to the Moorish power. The loss of the prince himself was the smallest part of the misfortune. Boabdil, though he could show true Moorish courage in the battle-field, was a weak and vacillating man, and was perpetually oppressed by the conviction that destiny was against him. He was known as Ez-Zogoiby, "the Unlucky;" and he was ever lamenting his evil star, against which he felt it was useless to struggle. "Verily," he would exclaim, after every reverse, "it was written in the book of fate that I should be unlucky, and that the kingdom should come to an end under my rule!" Boabdil could easily be spared; but innocuous as he was in himself, he might become dangerous in the hands of a clever adversary; and events showed that Boabdil's subjection to Ferdinand contributed as much as any other cause to the overthrow of the Moorish power in Andalusia. The Catholic sovereigns received him with honour at Cordova, and, by friendly persuasion and arguments drawn from his own desperate situation and the strongly contrasted successes of the Christians, they induced him to become their instrument and vassal.

As soon as they felt that they had completely mastered their tool, the politic king and queen suffered him to return to Granada, where his father, Abu-l-Hasan, once more held the fortress of the Alhambra. Favoured by his old supporters in the Albaycin quarter of the city, Boabdil managed to effect an entrance, and to seize the citadel or keep called Alcazaba, whence he carried on a guerilla warfare with his father in the opposite fort. The quarrel was further embittered by the rivalry between the wives of Abu-l-Hasan. Ayesha, the mother of Boabdil, was intensely jealous of a Christian lady, Zoraya, whom Abu-l-Hasan loved far beyond his other wives; and the chief courtiers took up the cause of either queen. Thus arose the celebrated antagonism between the Zegris, a Berber tribe from Aragon, who supported Ayesha, and the Abencerrages, or Beny-Serrāj, an old Cordovan family, which ended in the celebrated massacre of the Abencerrages in the Palace of Alhambra, though whether Boabdil was the author of this butchery is still matter of doubt. Supported by the Zegris, Boabdil for some time held his ground in the citadel. Old Abu-l-Hasan was too strong for him, however, and the son was soon compelled to take refuge at Almeria. Henceforward there were always two kings of Granada: Boabdil, on the one hand, always unlucky, whether in policy or battle, and despised by good Moors as the vassal of the common enemy; on the other, Abu-l-Hasan, or rather his brother Ez-Zaghal, "the Valiant," for the old king did not long survive the misfortunes which his son's rebellion had brought upon the kingdom. He lost his sight, and soon afterwards died, not without suspicion of foul play.

In Ez-Zaghal we see the last great Moorish King of Andalusia. He was a gallant warrior, a firm ruler, and a resolute opponent of the Christians. Had he been untrammelled by his nephew, Granada might have remained in the hands of the Moors during his life, though nothing could have prevented the final triumph of the Christians. Instead of delaying that victory, however, the kings of Granada did their best to further and promote it by their internal disputes. *Quem Deus vult perdere, prius dementat:* when the gods have decreed that a king must fall, they fill him first with folly. Such a suicidal mania now invaded the minds of the rulers of Granada: at a time when every man they could gather together was needed to repel the invasion of the Christians, they wasted their strength in ruinous struggles with each other, and one would even intercept the other's army when it was on the march against the common enemy. The people of Granada, divided into various factions, aided and abetted the jealousy of their sovereigns: always fickle and prone to any change, good or bad, the Granadinos loved nothing better than to set up and put down kings. So long as a ruler was fortunate in war, and brought back rich spoils from the territories of the "infidels," they were well pleased to submit to his sway; but the moment he failed, they shut the gates in his face and shouted, Long live the other!—who might be Boabdil or Ez-Zaghal, or any one else who happened for the moment to possess Granada's changeable affections.

While Boabdil the Unlucky was doing his best to

MOSQUE LAMP FROM GRANADA.

foil the efforts of his brave uncle Ez-Zaghal, the Christians were gradually narrowing the circle that they had drawn round the doomed kingdom. City after city fell into their hands. Alora and other forts were taken in 1484, with the aid of Ferdinand's heavy "lombards"—a new and destructive form of artillery. Coin, Cartama, Ronda, followed in the next year, not without some vigorous reprisals on the part of Ez-Zaghal, who caught the knights of Calatrava in an ambush, and effected a terrible slaughter. Still the course of Christian conquest steadily continued. Loxa fell in 1486, when an English Earl, Lord Scales, with a company of English archers, led the attack. Illora and Moclin succumbed; "the right eye of Granada is extinguished," cried the Moors in consternation; "the Catholic sovereigns have clipped the right wing of the Moorish vulture," was the Christian comment. The western part of the kingdom had, indeed, been absorbed by Ferdinand and his intrepid consort. The pomegranate (*granada*) was being devoured grain by grain. Ez-Zaghal became unpopular with the people, who could not brook disappointment, and they received Boabdil once more into their city. He found it hard work to maintain his foothold there against his uncle; but with the help of some troops furnished by the Christians he contrived to stand awhile at bay. Just then Ferdinand was laying siege to Velez, near Malaga, and the news roused the strongest feeling of indignation in Granada; for Malaga was the second city of the kingdom. Its site, shut in by mountains and the sea, its vineyards and orchards, gardens and pastures, and its fine defensive works, made it the right hand of the Moslem kingdom. If Malaga fell, then the Alhambra must also pass into the hands of the "eaters of swineflesh." Moved by the general emotion, and ever ready to break lance with the invader, Ez-Zaghal boldly led his troops to the relief of Velez. He knew that his treacherous nephew was in Granada, ready to take advantage of his absence to recover his old supremacy; but Ez-Zaghal was rightly called the Valiant; he put aside all thoughts of self, and set out to save Malaga. But he had to deal with a shrewd opponent; and while he took his measures for a combined attack from the besieged and the relieving army, Ferdinand intercepted his messages and countermined his plans. One night the people of Velez saw the hosts of Ez-Zaghal gathered in long array upon the neighbouring heights; the next morning not a soul remained; the night attack had failed, and the relieving army had melted like the mist before the resolute onslaught of the Marquess of Cadiz. When the dejected stragglers began to steal sadly into the gates of Granada, the populace easily threw off their old allegiance, and breaking into furious indignation against Ez-Zaghal, denounced him as a traitor, and proclaimed Boabdil king in his stead. As Ez-Zaghal drew near to the gates of Granada with the remnant of his army, he found them closed in his face, and looking up he saw the standard of Boabdil floating above the towers of the Alhambra. His city, always intolerant of failure, had shut its heart against him in his day of trouble, so he turned away and established his court at Guadix.

The siege of Malaga itself was now begun, but the strength of its defences rendered it a formidable obstacle. It was surrounded by mountains, defended by stout walls, overshadowed by the citadel and the still loftier Gibralfaro, or "Hill of the Beacon," whence its garrison could pour down missiles upon the Christians in the plain. Moreover, the defence was led by Ez-Zegry, an heroic Moor, who had been Alcayde of Ronda and could not forgive the Christians for wrenching that famous rocky fortress from him, and who now inspired the citizens and his following of African troops with a spirit of daring and endurance which the Catholic sovereigns in vain tried to subdue. Commanding the Gibralfaro, he was able to defend the city in spite of the peaceful inclinations of its trading classes. When the king attempted to bribe him, he dismissed the messenger with courteous disdain; and when the city was summoned to surrender, and the merchants eagerly acquiesced, Ez-Zegry said: "I was set here not to surrender but to defend." Ferdinand concentrated his attack upon the Gibralfaro; his terrible cannon, known as the "Seven Sisters of Ximenes," wrapped the castle in smoke and flame; night and day the artillery blazed to and fro. The Christians attempted to take the place by assault, but Ez-Zegry and his undaunted followers poured boiling pitch and rosin upon the assailants, hurled huge stones upon their heads as they climbed the ladders, and transfixed them with well-aimed arrows from the tower above, till the storming party were compelled to retire with heavy loss. Mines were tried with better success, and some of the fortifications were blown up with gunpowder, for the first time in Spanish history; but still the garrison held out. The chivalry of Spain was now gathered about the walls of Malaga; Queen Isabella herself came, and her presence infused a fresh spirit of enthusiasm into her knights and soldiers. Wooden towers were brought to bear upon the battlements; a *testudo* of shields was used as cover for the men who undermined the walls; but Ez-Zegry was still unsubdued. At last there appeared a worse enemy than cannon and gunpowder: famine began to distress the people of Malaga, and they were more inclined now to listen to the pacific policy of the traders than to the bold counsels of the commander. Help from without was not to be expected. Ez-Zaghal had, indeed, once more made an effort to save the besieged city. He had gathered together what was left of his army and gone forth from Guadix to succour Malaga; but his ill-starred nephew again proved his title to the name "Unlucky," for in a fit of insensate jealousy he ordered out the troops of Granada, intercepted Ez-Zaghal's small force as it was on its way to Malaga, and dispersed it. Ez-Zegry's last sally was repulsed with terrible slaughter; the people were starving, and mothers cast their infants before the governor's horse, lamenting that they had no more food and could not bear to hear their children's cries. The city at last surrendered, and Ez-Zegry, who still held out in the Gibralfaro, was forced by his soldiers to open the gates, and was rewarded for his heroism by being cast into a dungeon, never to be heard of again.

The long siege was over; the famished people

MALAGA.

fought with one another to buy food from the Christians. The African garrison, who still kept their proud look, though worn and enfeebled with their long struggle and privations, were condemned to slavery; the rest of the inhabitants were permitted to ransom themselves, but on these insidious terms—that all their goods should at once be paid over to the king as part payment, and that if after eight months the rest were not forthcoming, they should all be made slaves. They were numbered and searched, and then sent forth. "Then might be seen old men and helpless women and tender maidens, some of high birth and gentle condition, passing through the streets, heavily burdened, towards the Alcazaba. As they left their homes they smote their breasts, and wrung their hands, and raised their weeping eyes to heaven in anguish; and this is recorded as their plaint: O Malaga! city so renowned and beautiful, where now is the strength of thy castle, where the grandeur of thy towers? Of what avail have been thy mighty walls for the protection of thy children? . . . They will bewail each other in foreign lands; but their lamentations will be the scoff of the stranger." The poor people were sent to Seville, where they were kept in servitude till the eight months had expired, and then, since they had no money to pay the remainder of their ransoms, they were one and all condemned to perpetual slavery, to the number of fifteen thousand souls. Ferdinand's ungenerous ingenuity was thus rewarded.

The western part of the kingdom of Granada was now entirely in the hands of the Christians. The famous Moorish fortresses of the Serrania de Ronda and the beautiful city of Malaga held Christian garrisons. Granada itself was in the hands of Boabdil, who hastened to congratulate his liege lord and lady upon their triumph over Malaga. But in the east old Ez-Zaghal still turned a bold front to the invader, and gathered around his standard all that remained of patriotism among the disheartened Moors. From Jaen in the north, to Almeria, the chief port of Andalusia on the Mediterranean coast, his sway was undisputed; he held the important cities of Guadix and Baza; and within his dominion the rugged ridges of the Alpuxarras mountains, the cradle of a hardy and warlike race of mountaineers, sheltered countless valleys, fed with cool waters from the Sierra Nevada's snowy peaks, where flocks and herds, vines, oranges, pomegranates, citrons, and mulberry trees provided wealth for a whole province.

In 1488 Ferdinand turned his victorious arms towards this undisturbed portion of the Moorish dominion. Assembling his troops at Murcia, he marched westwards into Ez-Zaghal's territory, and attacked Baza. Here his advance was sternly checked; Ez-Zaghal's hand had not lost its ancient cunning, and he drove the Christians back from the walls of Baza, and began to retaliate by making raids into their own country. In the following year Ferdinand, nothing disheartened, renewed his attack on Baza; but instead of sacrificing his troops in vain assaults, he laid waste the fertile country round about, and so starved the city into submission. It took six months, and the Christians lost twenty thousand men from disease and exposure, joined to the accidents of war; but in December, 1489, Baza finally submitted, and with the loss of this chief city Ez-Zaghal's power was broken. The castles that dominated the fastnesses of the Alpuxarras yielded one by one to Ferdinand's prestige or gold. Ez-Zaghal perceived that the rule of the Moors was doomed: reluctantly he gave in his submission to Ferdinand, and surrendered the city of Almeria. He was allotted a small territory in the Alpuxarras, with the title of King of Andarax. He did not long remain in the land of his lost glory and present shame; he sold his lands and went to Africa, where he was cruelly blinded by the Sultan of Fez, and passed the remainder of his days in misery and destitution, a wandering outcast,—pitied by those who could recognize the hero in a mendicant's rags, or read the badge which he wore, whereon was written in the Arabic character, "This is the hapless King of Andalusia."

Granada alone remained to the Moors. Boabdil had been well pleased to see his old rival Ez-Zaghal dethroned by their Catholic Majesties: "Henceforth,' he cried to the messenger who brought him the news, "let no man call me Zogoiby, for my luck has turned:" to which the other made answer that the wind which blew in one quarter might soon blow in another, and the king had best reserve his rejoicings for more settled weather. Boabdil, though he heard his name cursed in the streets of his capital as a traitor in league with the infidels, indulged in blind confidence, now that his detested uncle was powerless; as the vassal of Ferdinand and Isabella he believed that he had nothing to fear. He had forgotten that when, in his fatuous hatred of Ez-Zaghal, he incited the Christian sovereigns to subdue his rival's dominions, he had engaged by treaty that should Ferdinand succeed in reducing Ez-Zaghal's country, with the cities of Guadix and Almeria, he would on his part surrender Granada. He was not, however, long left without a spur to his memory. Ferdinand wrote to inform him that the conditions named in the treaty had been fulfilled on his side, and demanded the surrender of Granada in accordance with the terms then laid down. Boabdil in vain implored delay; the king was determined, and threatened to repeat the example of Malaga if the capital were not immediately given up. Boabdil did not know what to reply; but the people of Granada, led by Mūsa, a brave and gallant knight, took the matter into their own hands, and told his Catholic Majesty that if he wanted their arms he must come and take them!

When these bold words were said, the beautiful Vega of Granada was waving with crops and fruit; it had recovered from the devastations which accompanied the struggle between Ez-Zaghal and Boabdil, and a splendid harvest was awaiting the sickle. Ferdinand saw his opportunity, and, adopting his usual tactics, poured his troops, twenty-five thousand strong, over the Vega, and for thirty days abandoned it to their destroying hands. When he turned back towards Cordova, the Vega was one great expanse of desolation. It was enough for one season; yet once more was the cruel work of destruction carried out in that year of grace 1490.

SWORD OF BOABDII. (*Villaseca Collection, Madria*).

Boabdil had at last been roused to a desperate courage. Guided by Mūsa, whose mettle was of the finest, he girded on his armour, and began to carry the war into the enemy's quarters. The Moors round about, who had given in their submission to Ferdinand, were heartened by the sight of the King of Granada once more on the war path, and, hastily consigning their promises to the winds, rose up and joined him. It really seemed as if the good old days of Granada were returning; some fortresses were recovered from the Christians, and the Moorish army ravaged the borders. It was but the last gleam of light before the final setting of the sun. In April, 1491, Ferdinand and Isabella set forth upon their annual crusade, resolved not to return till Granada was in their power. The king led an army of forty thousand foot and ten thousand horse, with such commanders as the famous Ponce de Leon, Marquess of Cadiz, the Marquess of Santiago, the Counts of Tendilla and Cabra, the Marquess of Villena, and the redoubtable knight, Don Alonzo de Aguilar. Boabdil held a council in the Alhambra, whence the clouds of dust raised by Christian horsemen could be seen on the Vega; some urged the futility of resistance, but Mūsa got up and bade them be true to their ancestors and never despair while they had strong arms to fight and fleet horses wherewith to foray. The people caught Mūsa's enthusiasm, and there was nothing heard in Granada but the sound of the furbishing of arms and the tramp of troops.

Mūsa was in chief command, and the gates were in his charge. They had been barred when the Christians came in view; but Mūsa threw them open. "Our bodies," he said, "will bar the gates." The young men were kindled by such words, and when he told them, "We have nothing to fight for but the ground we stand on; without that we are without home or country," they made ready to die with him. With such a leader, the Moorish cavaliers performed prodigious feats of valour in the plain which divided the city from the Christian camp. Single combats were of daily occurrence; the Moors would ride almost among the tents of the Spaniards, and tempt some knight to the duel, from which he too often did not return. Ferdinand found his best warriors were being killed one by one, and he straitly forbade his knights to accept the Moors' challenge. It was hard for the Spanish chivalry to sit still within their tents, while a bold Moorish horseman would ride within hail and taunt them with cowardice; and when at length one of the Granadinos waxed so venturesome that he cast a spear almost into the royal pavilion, Hernando Perez de Pulgar, surnamed "He of the Exploits," could no longer contain himself, but gathering a small band of followers, rode in the dead of night to a postern gate in the walls of Granada, and, surprising the guards, galloped through the streets till he came to the chief mosque, which he forthwith solemnly dedicated to the Blessed Virgin, and in token of its conversion nailed a label on the door inscribed with the words *Ave Maria.* Granada was awake by this time, and soldiers were gathering in every direction; but Pulgar put spurs to his horse, and, amid the amazement of the people, plunged furiously through the crowd, overturning them as he galloped to the gate, and, fighting his way out, rode back in triumph to the camp. The Pulgars ever after held the right to sit in the choir of the mosque-church during the celebration of High Mass.

Such feats of daring, however, did little to advance the siege, nor were the few engagements conclusive. Ferdinand renewed his old tactics. He sallied forth from his camp, which had accidentally been burnt to the ground, and proceeded to lay waste what remained of the fertility of the Vega. The Moors made a last desperate sally to save their fields and orchards, and Mūsa and Boabdil fought like heroes at the head of their cavalry; but the foot soldiers, less steadfast, were beaten back to the gates, whither Mūsa sadly followed them, resolved never again to risk a pitched battle with such men behind him. It was the last fight of the Granadinos. For ten years they had disputed every inch of ground with their invaders; wherever their feet could hold they had stood firm against the enemy. But now there was left to them nothing beyond their capital, and within its walls they shut themselves up in sullen despair. To starve them out was an agreeable task for the Catholic king; and following the precedent of the third Abd-er-Rahmān in the siege of Toledo, he built in eighty days a besieging city over against Granada, and called it Santa Fé, in honour of his "Holy Faith," and there to this day it stands, a monument of Ferdinand's resolution. Famine did the work that no mere valour could effect. The people of Granada implored Boabdil to spare them further torture and make terms with the besiegers, and at last the unlucky king gave way. Mūsa would be no party to the surrender. He armed himself *cap-à-pie*, and mounting his charger rode forth from the city never to return. It is said that as he rode he encountered a party of Christian knights, half a score strong, and, answering their challenge, slew many of them before he was unhorsed, and then, disdaining their offers of mercy, fought stubbornly upon his knees, till he was too weak to continue the struggle: then with a last effort he cast himself into the river Xenil, and, heavy with armour, sank to the bottom.

On the 25th of November, 1491, the act of capitulation was signed, and a term was fixed during which a truce was to be observed, after which, should no aid come from outside, Granada was to be delivered up to their Catholic Majesties. In vain the Moors watched for a sign of the help they had sought from the Sultans of Turkey and Egypt. No aid came, and at the end of December Boabdil sent a message to Ferdinand to come and take possession of the city. The Christian army filed out of Santa Fé, and advanced across the Vega, watched with mournful eyes by the unhappy Moors. The leading detachment entered the Alhambra, and presently the great silver cross was seen shining from the summit of the Torre de la Vela; beside it floated the banner of St. James, while shouts of "Santiago!" rose from the army in the plain beneath; and lastly, the standard of Castile and Aragon was planted by the side of the cross. Ferdinand and Isabella fell on their knees and gave thanks to God; the whole army of Spain knelt behind them, and the royal choir sang a solemn *Te Deum*. At the foot of the Hill of Martyrs, Boabdil, attended by a small band of horsemen, met the royal procession. He gave Ferdinand the keys of Granada, and, turning his back upon his beloved city, passed on to the mountains. There, at Padul, on a spur of the Alpuxarras, Boabdil stood and gazed back upon the kingdom he had lost: the beautiful Vega, the towers of Alhambra, and the gardens of the Generalife; all the beauty and magnificence of his lost home. "Allahu Akbar," he said, "God is most great," as he burst into tears. His mother Ayesha stood beside him: "You may well weep like a woman," she said, "for what you could not defend like a man." The spot whence Boabdil took his sad farewell look at his city from which he was banished for ever, bears to this day the name of *el ultimo sospiro del Moro*, "the last sigh of the Moor." He soon crossed over to Africa, where his descendants learned to beg their daily bread.

There was crying in Granada when the sun was going down;
Some calling on the Trinity—some calling on Mahoun.
Here passed away the Koran—there in the Cross was borne—
And here was heard the Christian bell—and there the Moorish horn:

Te Deum Laudamus! was up the Alcala sung:
Down from the Alhambra's minarets were all the crescents flung;
The arms thereon of Aragon they with Castile display;
One king comes in in triumph—one weeping goes away.

Thus cried the weeper, while his hands his old white beard did tear,
Farewell, farewell, Granada! thou city without peer!
Woe, woe, thou pride of heathendom! seven hundred years and more
Have gone since first the faithful thy royal sceptre bore!

Thou wert the happy mother of a high renownèd race;
Within thee dwelt a haughty line that now go from their place;
Within thee fearless knights did dwell, who fought with mickle glee,
The enemies of proud Castile, the bane of Christentie.

Here gallants held it little thing for ladies' sake to die,
Or for the Prophet's honour, and pride of Soldanry;
For here did valour flourish and deeds of warlike might
Ennobled lordly palaces in which was our delight.

The gardens of thy Vega, its fields and blooming bowers—
Woe, woe! I see their beauty gone, and scattered all their flowers!
No reverence can he claim—the king that such a land hath lost—
On charger never can he ride, nor be heard among the host;
But in some dark and dismal place, where none his face may see,
There weeping and lamenting, alone that king should be.[1]

[1] Lockhart: Spanish Ballads.

XIV.

BEARING THE CROSS.

BOABDIL'S "last sigh" was but the beginning of a long period of mourning and lamentation for the luckless Moors he had ushered to destruction. At first, indeed, it seemed as if the equitable terms upon which Granada had capitulated would be observed, and freedom of worship and the Mohammedan law would be upheld. The first archbishop, Hernando de Talavera, was a good and liberal-minded man, and forcible conversion formed no part of his policy. He strictly respected the rights of the Moors, and sought to win them over by force of example, by uniform justice and kindness, and by conforming as far as possible to their ways. He made his priests learn Arabic, and said his prayers in the same ungodly tongue, and by such concessions "so wrought on the minds of the populace that in 1499, when Cardinal Ximenes was sent by the queen to aid him in the work, it seemed as if the scenes which occurred at Jerusalem in the infancy of the Faith were about to be reënacted at Granada. In one day no less than 3,000 persons received baptism at the hands of the Primate, who sprinkled them with the hyssop of collective regeneration."[1] Ximenes was little in harmony

[1] Sir W. Stirling Maxwell: Don John of Austria, i. 115.

with the archbishop's soft ways: he was the apostle of the Church Militant, always most active when militant meant triumphant, and would have the souls of these "infidels" saved from hell fire whether they liked it or no. He insinuated in Isabella's holy mind the pernicious doctrine that to keep faith with infidels was breaking faith with God; and it is one of the few blots on the good queen's name that she at length consented to the persecution of the Moors—or "Moriscos," as they now began to be called.

The first attempt to coerce the Granadinos was a failure. Some of the straiter Moslems expressed their repugnance to the new conversions to Christianity, and these malcontents were arrested. A woman being haled to prison on such a pretext roused the people of the Albaycin; they rose in arms and rescued her, and Granada was filled with uproar and barricade-fights. The garrison was hopelessly outnumbered; Ximenes raged with impotent fury; but the peaceful archbishop went forth, followed only by his cross-bearer, and, fearlessly entering the Albaycin, was at once surrounded by the people, who kissed his garments, and laid their wrongs before him in whom they accepted a just and generous mediator. Talavera composed the disputes, and the Cardinal had to retire.

Ximenes was, however, not a man to be easily deterred from his purpose. He induced the queen to promulgate a decree by which the Moors were given their choice of baptism or exile. They were reminded that their ancestors had once been Christian, and that by descent they themselves were born in the

Church, and must naturally profess her doctrine. The mosques were closed, the countless manuscripts that contained the results of ages of Moorish learning were burnt by the ruthless Cardinal, and the unhappy "infidels" were threatened and beaten into the Gospel of Peace and Goodwill after the manner already approved by their Catholic Majesties in respect of the no less miserable Jews. The majority of course yielded, finding it easier to spare their religion than their homes; but a spark of the old Moorish spirit remained burning bright among the hillmen of the Alpuxarras, who for some time held their snowy fastnesses against their persecutors. The first effort to suppress the rebellion ended in disaster. Don Alonzo de Aguilar, whose fame in deeds of derring-do had been growing for forty years of valiant chivalry, was sent into the Sierra Bermeja in 1501, and sustained a terrible defeat at the hands of the Moriscos, who crushed his cavalry with the massive rocks which they hurled down upon them.

Beyond the sands, between the rocks, where the old cork trees grow,
The path is rough, and mounted men must singly march and slow;
There o'er the path the heathen range their ambuscado's line,
High up they wait for Aguilar, as the day begins to shine.

There naught avails the Eagle eye, the guardian of Castile,
The eye of wisdom, nor the heart that fear might never feel,
The arm of strength that wielded well the strong mace in the fray,
Nor the broad plate from whence the edge of falchion glanced away.

Not knightly valour there avails, nor skill of horse and spear;
For rock on rock comes rumbling down from cliff and cavern drear;
Down, down like driving hail they come, and horse and horseman die
Like cattle whose despair is dumb when the fierce lightnings fly.

Alonzo with a handful more escapes into the field,
There like a lion, stands at bay, in vain besought to yield;
A thousand foes around are seen, but none draws near to fight,
Afar with bolt and javelin, they pierce the steadfast knight.

A hundred and a hundred darts are hissing round his head;
Had Aguilar a thousand hearts, their blood had all been shed;
Faint and more faint he staggers upon the slippery sod—
At last his back is to the earth, he gives his soul to God.

Another and more probable legend, however, tells how Aguilar was killed in fair fight by the commander of the Moors. He was the fifth lord of his line who died in combat with the infidels.

This temporary success, however, only aggravated the reprisals of the now exasperated Christians. The Count of Tendilla stormed Guejar; the Count of Serin "blew up the mosque in which the women and children of a wide district had been placed for safety," and King Ferdinand himself seized the key of the passes, the castle of Lanjaron. The remnant of the rebels fled to Morocco, Egypt, and Turkey, where their skill as artificers secured them a living. Thus the first revolt in the Alpuxarras was suppressed.

Half a century of smouldering hatred ensued. The Moriscos grudgingly fulfilled the minimum of the religious duties imposed on them by their outward conversion; but they took care to wash off the holy water with which their children were baptized as soon as they were out of the priest's sight; they came home from their Christian weddings to be married again after the Mohammedan rite; and they made the Barbary corsair at home in their cities, and helped him to kidnap the children of the Christians. A wise and honest government, respecting its pledges given at the surrender of Granada, would have been spared the dangers of this hidden disaffection ; but the rulers of Spain were neither wise nor honest in their dealings with the Moriscos, and as time went on they became more and more cruel and false. The "infidels" were ordered to abandon their native and picturesque costume, and to assume the hats and breeches of the Christians ; to give up bathing, and adopt the dirt of their conquerors ; to renounce their language, their customs and ceremonies, even their very names, and to speak Spanish, behave Spanishly, and re-name themselves Spaniards. The great Emperor Charles V. sanctioned this monstrous decree in 1526, but he had the sense not to enforce it ; and his agents used it only as a means of extorting bribes from the richer Moors as the price of official blindness. The Inquisition was satisfied for the time with a "traffic in toleration" which filled the treasury in a highly satisfactory way. It was reserved for Philip II. to carry into practical effect the tyrannical law which his father had prudently left alone. In 1567 he enforced the odious regulations about language, customs, and the like, and, to secure the validity of the prohibition of cleanliness, began by pulling down the beautiful baths of the Alhambra. The wholesale denationalization of the people was more than any folk—much less the descendants of the Almanzors, the Abd-er-Rahmāns, and the Abencerrages—could stomach. A fracas with some plundering tax-gatherers set light to the inflammable materials which had long been ready to burn up : some soldiers were murdered by peasants in whose huts they were billeted ; a dyer of Granada,

Farax Aben Farax, of the blood of the Abencerrages, gathered together a band of the disaffected, and escaped to the mountains before the garrison had made up their minds to pursue him; Hernando de Valor, of the race of the Khalifs of Cordova, a man of note in Granada, but brought to disgrace by his dissolute habits, was chosen King of Andalusia, with the title of Muley Mohammed Aben Omeyya; and in a week the whole of the Alpuxarras was in arms, and the second Morisco rebellion had begun (1568).

The district of the Alpuxarras was well fitted to harbour a revolt. The stretch of high land between the Sierra Nevada and the sea, about nineteen miles long and eleven broad, is "so rudely broken into rugged hill and deep ravine, that it would be hard to find in its whole surface a piece of level ground, except in the small valley of Andarax and on the belt of plain which intervenes betwixt the mountains and the sea. Three principal ranges, spurs of the Sierra Nevada, and themselves spurred with lesser offshoots, intersect it from north to south. Through the glens thus formed a number of streams—torrents in winter but often dry in summer—pour the snows of Muleyhacen and the Pic de Valeta into the Mediterranean. In natural beauty, and in many physical advantages, this mountain land is one of the most lovely and delightful regions of Europe. From the tropical heat and luxuriance, the sugar-canes and the palm-trees, of the lower valleys and of the narrow plain which skirts the sea like a golden zone, it is but a step, through gardens, steep cornfields, and olive groves, to fresh Alpine pastures and woods of pine, above which vegetation expires on the rocks where snow lies long and deep, and is still found in nooks and hollows in the burning days of autumn. When thickly peopled with laborious Moors, the narrow glens, bottomed with rich soil, were terraced and irrigated with a careful industry which compensated for want of space.[1] The villages, each nestling in its hollow, or perched on a craggy height, were surrounded by vineyards and gardens, orange and almond orchards, and plantations of olive and mulberry, hedged with the cactus and aloe; above, on the rocky uplands, were heard the bells of sheep and kine; and the wine and fruit, the silk and oil, the cheese and the wool of the Alpuxarras, were famous in the markets of Granada and the seaports of Andalusia."[2] It was this beautiful province that the bigotry of the priest was about to deliver over to the sword and brand of the soldier.

The great rebellion in the Alpuxarras lasted for two years, and its repression called forth the utmost energy of the Spaniards. Its records are full of deeds of reckless bloodshed, of torture, assassination, treachery, and horrible brutality on both sides; but they are relieved by acts of heroism and endurance which would do honour to any age and any nation. The struggle was fierce and desperate: it was the Moors' last stand; they felt themselves at bay, and

[1] The Spaniards were never able to do justice to the rich soil of Andalusia. So little did the Crown think of the fertile country about Granada that in 1591 the royal domains there were sold, because they cost more than the Spaniards could make them yield! In the time of the Moors the same lands were gardens of almost tropical luxuriance.

[2] Sir W. Stirling Maxwell: Don John of Austria, i. 126–8.

they avenged in their first mad rush of fury a hundred years of insult and persecution. Village after village rose against its oppressors; churches were desecrated, Our Lady's picture was made a target, priests were murdered, and too often horrid torture was used against the Christians, who, for their part, took refuge in belfries and towers, and valiantly resisted the sudden assault of the enemy. We read how two women, left alone in a tower, fastened the door, and armed only with stones which they aimed from the battlements, wounded by arrows, and supported by nothing save their own brave hearts, kept out their assailants from dawn till noon, when relief fortunately came. Another golden deed is told of the advance of the Christian expedition to put down the revolt. The troops had arrived at the ravine of Tablete, a grim chasm, a hundred feet deep, with a roaring torrent at the bottom. The Moriscos had destroyed the bridge, and only a few tottering planks remained, by which a venturesome scout might cross if needful. On the other side of these planks Moorish archers kept their bows at stretch. It is not surprising that the soldiers recoiled from such a crossing; the dancing plank, the torrent's roar, and the Moorish arrows, were enough to daunt the bravest. While the army stood irresolute, a friar came to the front, and calmly led the way across the plank over the torrent, to the very arrows of the enemy, who were too much struck with admiration to think of shooting. Two soldiers sprang after the devoted friar—one reached the other side, the other fell into the hissing flood beneath. Then the whole army plucked up heart and crossing as quickly as they could, and mustering on the other side, charged up the slope, and carried the position. It was a Thermopylæ reversed, with a friar for its Leonidas; a Balaclava galloped upon quicksands; and it redeems a long catalogue of baseness.

The Marquess of Mondéjar, who commanded at Granada, endeavoured by conciliation and generosity to calm the rebellion, which his resolute march into the mountains at the head of four thousand men had to a great extent suppressed; but an accidental massacre at Jubiles, and an act of treachery at Laroles, rekindled the flame of revolt which had been partly extinguished; and the ruthless murder of one hundred and ten Moriscos by their Christian fellow-prisoners in the jail of the Albaycin still further exasperated the persecuted race. Mondéjar was innocent of any share in this bloody work, and was marching with his guard to the prison to quell the disturbance, when the Alcayde met him with the remark: "It is unnecessary; the prison is quiet—*the Moors are all dead.*" After this the Moriscos gained daily in strength, and Aben Umeyya became really lord of the whole district of the Alpuxarras. This incapable and profligate sprig of Cordovan nobility enjoyed his power for a very brief period, however; for in October, 1569, private spite and suspicion led to his being strangled in bed by his own followers, when an able and devoted man, the true leader of the rebellion, and one who could even dare to die for his friend, assumed the title of king as Muley Abdallah Aben Abó.

Aben Abó had to deal with a new opponent. The king's half-brother, Don John of Austria, a young man of twenty-two, but full of promise, superseded Mondéjar as commander-in-chief against the Moriscos, and after a protracted war of letters he convinced Philip of the gravity of the situation and the necessity for strong measures. At last Don John received his marching orders, and after that, it was but a short shrive that the Moriscos had to expect. In the winter of 1569–70 he began his campaign, and in May the terms of surrender had been arranged. The months between had been stained with a crimson river of blood. Don John's motto was "no quarter"; men, women, and children were butchered by his order and under his own eye; the villages of the Alpuxarras were turned into human shambles.

Even when the rebellion seemed at an end, a last feeble flicker of revolt once more sprang up: Aben Abó was not yet reconciled to oppression. Assassination, however, finally convinced him: his head was exhibited over the Gate of the Shambles at Granada for thirty years. The Grand Commander, Requesens, by an organized system of wholesale butchery and devastation, by burning down villages, and smoking the people to death in the caves where they had sought refuge, extinguished the last spark of open revolt before the 5th of November, 1570. The Moriscos were at last subdued, at the cost of the honour, and with the loss of the future, of Christian Spain.

Slavery and exile awaited the survivors of the rebellion. They were not very many. The late wars, it was said, had carried off more than twenty thousand Moors, and perhaps fifty thousand remained in the district on that famous Day of All Saints, 1570, when the honour of the apostles and martyrs of Christendom was celebrated by the virtual martyrdom of the poor remnant of the Moors. Those taken in open revolt were enslaved, the rest were marched away into banishment under escort of troops, while the passes of the hills were securely guarded. Many hapless exiles died by the way, from want, fatigue, and exposure; others reached Africa, where they might beg a daily pittance, but could find no soil to till; or France, where they received a cool welcome, though Henry IV. had found them useful instruments for his intrigues in Spain. The deportation was not finished till 1610, when half a million of Moriscos were exiled and ruined. It is stated that no less than three million of Moors were banished between the fall of Granada and the first decade of the 17th century. The Arab chronicler mournfully records the *coup-de-grâce*: "The Almighty was not pleased to grant them victory, so they were overcome and slain on all sides, till at last they were driven forth from the land of Andalusia, the which calamity came to pass in our own days, in the year of the Flight, 1017. Verily to God belong lands and dominions, and He giveth them to whom He doth will."

The misguided Spaniards knew not what they were doing. The exile of the Moors delighted them; nothing more picturesque and romantic had occurred for some time. Lope de Vega sang about the *sentencia justa* by which Philip III., *despreciando sus barbaros tesoros*, banished to Africa *las ultimas reliquias de los Moros;* Velazquez painted it in a memorial picture; even the mild and tolerant Cervantes forced himself to justify it. They did not understand that they had killed their golden goose. For centuries Spain had been the centre of civilization, the seat of arts and sciences, of learning, and every form of refined enlightenment. No other country in Europe had so far approached the cultivated dominion of the Moors. The brief brilliancy of Ferdinand and Isabella, and of the empire of Charles V., could found no such enduring preëminence. The Moors were banished; for a while Christian Spain shone, like the moon, with a borrowed light; then came the eclipse, and in that darkness Spain has grovelled ever since. The true memorial of the Moors is seen in desolate tracts of utter barrenness, where once the Moslem grew luxuriant vines and olives and yellow ears of corn; in a stupid, ignorant population where once wit and learning flourished; in the general stagnation and degradation of a people which has hopelessly fallen in the scale of the nations, and has deserved its humiliation.

INDEX TO THE TEXT AND THE NOTES.

A

Abbadites, 176
Abbāside, 59, 60, 63–4
Abdallah, 98–107
Abd-el-Melik, 55, 56
Abd-er-Rahmān I., 33, 57, 59–68, 131, 136
Abd-er-Rahmān II., 78–94
Abd-er-Rahmān III., 107–128
Abd-er-Rahmān of Narbonne, 28
Aben Abó, 277–8
Abencerrages, 227, 247
Aben Dmeyya, 274
Abu-l-Hasan (Alboacen), 232 *ff.*, 247
Acisclus, St., 89
Aguilar, Don Alonzo de, 237, 271–2
Ahmar, Ibn-el-, 218
Alans, 6
Alarcos, 217
Albarracin, 209
Albaycin, 247, 271, 277
Albucasis, 144
Alcazar of Cordova, 131
Aledo, 177, 180
Alexander the Great, 1
Alexandria, 76
Alferez, 240
Alfonso I., 33
Alfonso IV., 176–181, 186, 194–196, 206
Alfonso the Battler, 184
Alfonso the Learned, 194, 218
Algarve, 110
Algeciras, 13, 179, 214, 221
Alhama, 235
Alhambra, 221 *ff.*
Alhandega, 123
Almanzor, 156–166
Almeria, 148, 151, 176
Almohades, 214
Almoravides, 178–184
Alpuente, 209
Alpuxarras, 259, 271–280
Alvar Fañez, 181, 196
Alvaro, 86
Amir, Ibn-Aby-, 156–166
Andalus, Emir of, 51
Andalusia, 43
Andarax, 259
Antequera, 236
Aquitaine, 28, 29
Arabic Studies, 90
Arabs, pre-Mohammedan, 1
Aragon, 208, 218
Archidona, 25, 62
Arts in Andalusia, 147
Asturias, 27, 33, 35, 116 *ff.*, 186
Aurora, 156, 157, 158, 161, 164
Avenzoar, 144
Averroes, 144
Axarquia, 237
Ayesha, 225, 247

B

Badajoz, 119, 179, 186, 217
Barcelona, 165, 166, 201

Basques, 13, 34
Bavieca, 210, 213
Baza, 258, 259
Beaune, 28
Bedr, 61
Beja, 63
Bellido, 195
Berbers, 4, 13*n*., 20, 40, 52–6, 65, 101, 109, 167–184
Bermudez, Pero, 201, 213
Bernardo del Carpio, 34
Beytar, Ibn-, 144
Boabdil, 225, 242, 245, 246 *ff*., 267
Bobastro, 102, 110
Body-guard, 66, 75, 114, 158
Bordeaux, 29
Burgos, 197
Burgundy, 28

C

Cabra, Count of, 242, 263
Cadiz, 177–8, 184
Cadiz, Marquess of, 235, 236, 238, 252, 263
Calahorra, 206
Calatrava, 251
Campeador, 192, 195
Carcasfonne, 28
Cardeña, St. Pedro de, 199, 213
Carmona, 28, 63, 184
Castile, 123 *ff*., 165, 189
Cava, 11 *n*.
Cazlona, 105
Ceuta, 4, 54, 55, 217
Cid, The, 177, 178, 181, 191–213
Charlemagne, 30, 33-8, 57, 65
Charles V., 222, 225, 231, 273
Charles Martel, 29, 30
Christian disaffection, 83 *ff*.
Christian power, 116 *ff*., 185 *ff*.
Christianity in Roman and Gothic Spain, 6–8
Chronicle of the Cid, 192, 195 *ff*.
Coimbra, 186
Cordova, 24, 26, 62, 74, 78, 106–7, 129–145, 184, 218
Coria, 55
Covadonga, 116–7

D

Darro, 225
Dhu-n-Nūn, 101, 176
Dozy, 47, 52, 56, 63, 76, 122, 127, 163, 176, 192
Durenda, 36–7

E

Elvira, 25, 56, 102
Emir, 121
Estevan de Gormaz, San, 119, 120
Estramadura, 101
Eudes, 28, 29, 55
Eulogius, 86–95

F

Fakis, 76
Farax, 274
Fātimite Khalifs, 115
Ferdinand and Isabella, 232, 251, 257, 260 *ff*.
Fernando I. of Leon and Castile, 186
Fernando III., 218
Feth, El-, 113
Fez, 76
Flora, 86–93
Florinda, 11
Foss, Day of the, 74
France, Arab advance into, 28–30
Franks, 29

G

Galicia, 55, 118, 165, 186
Garcia, 123
Garonne, 29
Gayangos, 56*n*
Gebal-Tārik (Gibraltar), 14
Generalifé, 228, 231
Gerona, 148
Ghālib, 159
Gibralfaro, 253, 254
Gonzalez, Fernando, 123–5
Goths, 4–8, 26
Granada, 25, 102, 184, 217 *ff*., 267

Greek ambassadors, 143
Greek Empire, 3, 4
Guadalete, 14, 23
Guadarrama, 40, 185
Guadelquivir, 40, 131, 135
Guadix, 252, 254, 258
Guarinos, 35

H

Hafsūn, Ibn-, 102, 106, 107, 110
Hajjāj, Ibn-, 105–6
Hakam I., 74–7, 78
Hakam II., 152–6
—— his library, 155
Hamdin, Ibn-, 184
Hammūd, 175, 176
Harūn-er-Rashīd, 78, 81
Hasdai, 125–6
Hayyān, Ibn-, 67, 116
Henry VI., 279
Hishām I., 71–4, 136
Hishām II., 156–171
Hishām III., 171
Hroswitha, 144
Hūd, Ibn-, 217

I

Isaac the monk, 88, 89
Isaac the Mosilite, 81
Isabella, 232, 251. 254, 260, 269
Isidore of Beja. 48
Islam, 2
Irving, Washington, 19, 221 *ff.*, 232 *ff.*

J

Jaen, 56
Jayme I., 218
Jews of Spain, 24
John of Austria, Don. 278
Julian, 4, 11, 12, 13, 27

K

Kādy, 87
Kāsy, Ibn-, 184
Khalif, 23, 27, 51, 56, 58–60
Khalif of Spain, 122
Kharaj, 44

L

Lamtūny, 184
Lanjaron, 272
Laroles, 277
Leon, 34, 35, 118, 159, 163, 189
Leon chivalry, 119. 190
Library of Hakam, 155
Lockhart, 21, 34–5, 124, 267, 271
Lorca, 101
Lormego, 186
Lothair, 29
Louis the Debonnaire, 83
Loxa, 251
Lucena, 242

M

Majolica, 148
Makkary, 56*n*, 128, 131
Malaga, 25, 56, 214, 251, 257
Malaga, the mountains of, 236 *ff.*
Mamlūks, 114
Mansur, the Khalif, 64
Marabout, 53
Mardanish, Ibn-, 184
Martin, Abbey of St., 29
Mary, 92–3
Maxwell, Sir W. Stirling, 269, 275
Maymūn, Ibn-, 184
Medina, 73
Medinaceli, 166
Merida, 28, 47, 55, 119
Mohammed I., 94, 98
Mohammed the Prophet, 2
Mahammedanism, 2
Mohammedan conquests, 3
Mondéjar, Marquess of, 277
Monousa, 55
Moor, 13*n*
Moriscos, 270 *ff.*
Mosque of Cordova, 136 *ff.*
Mo'temid, 176, 178, 180
Muez, 121
Mughīth, 23, 24
Mughīth, Ibn-, 63
Mundhir, 98
Murcia, 25, 103. 110, 176, 218
Murviedro, 209
Mūsa of Granada, 263–6
Mūsa, son of Noseyr, 12, 13, 23, 27, 28

Mushafy, 158–160
Mutes, 75, 76
Muzaffar, 169

N

Najera, 206
Narbonne, 28, 30, 136
Nāsir-li-dīni-llāh, En-, 122
Nasr, 81, 89
Nasr, Beny-, 217 *ff.*
Navarre, 119–121, 165, 166
Navas, Las, 217

O

Oliver, 37
Omeyyads, 33, 57, 59, 60, 62 *f.*
Ordoño II., 119, 120, 121
Ordoño, IV., 125
Orelia, 19
Orihuela, 25, 47
Osma, 119, 120
Ostrogoths, 4

P

Paderborn, 33
Padul, 267
Pamplona, 166
Pavement of Martyrs, 30
Pelagius, or Pelayo, 33, 116–7
Perfectus, 89
Philip II., 273
Philip III., 279
Pinos, 226
Poictiers, 29
Pulgar, 264

Q

Quixote, Don, 35

R

Ramiro II., 122
Regio, 110
Renegades, 48. 102
Requesens, 278
Roland, 36–8
Roderick, 4, 8. 11–22, 48
Roderick's vision, 18, 19
Roncesvalles, 34–8, 65
Ronda, 251, 258

S

Sacralias, 179
Sancho, 90
Sancho of Navarre, 119–121
Sancho of Castile, 195
Sancho the Fat, 125
Santa Fé, 265
Santiago, Master of, 238 *ff.*
Santiago de Compostella, 165
Saracens, 3
Science, 147
Seddaray, 184
Septimania, 28
Seville, 28, 62, 105, 109, 170–1, 176, 180, 184, 186, 214, 218
Sierra Nevada, 274
Simancas, 119
Slaves, 48
Slavs, 114, 158, 161, 170, 171, 175
Southey's Cid, 193
Spain under the Romans and the Goths, 4, 5–8
Suevi, 4, 6
Sultān, 121

T

Tablete, 276
Talavera, Archbishop, 269, 270
Tarīf, 13
Tarīfa, 13, 181
Tārik, 13, 20, 21, 23–28
Tarraconaise, 29
Tarūb, 81
Taxes, 44
Tendilla, Count of, 235, 263
Theodemir of Murcia, 25
Theological students, 73–6, 161
Theuda, 123
Tizona, 213
Toledo, 12, 14, 26, 28, 64, 74, 94, 102, 110, 148, 176, 186
Toledo, enchanted tower, 14–19
Toulouse, 28
Tours, 29, 30

Tribes, Arab, 50–2, 56, 101
Tudela, 120
Turpin, pseudo-, 35

V

Val de Junqueras, 120
Valencia, 176, 178, 182, 184, 205–213, 218
Vandals, 6
Vega, 221, 260
Velez, 251
Visen, 186
Visigoths, 4–8

W

Wady Bekka, 14
Welīd the Khalif, 23, 28
Wittekind, 33, 34
Witiza, 8, 11, 20, 21, 27

X

Xativa, 205
Xeres, 184
Ximena, 198, 199, 200, 213
Ximenes, Cardinal, 269, 270
Ximenes, Seven Sisters of, 253

Y

Yahyā, 73
Yahyā of Valencia, 205
Yemen tribes, 61, 65
Yūsuf the Almoravide, 179–181
Yūsuf, 62

Z

Zāb, Prince of, 164
Zaghal, Ez-, 240, 247 *ff.*, 259
Zahara, 232–4
Zahrā, Medinat-ez-, 140–4, 175
Zallāka, 179
Zamora, 119, 195
Zaragoza, 34, 65, 101, 122–3, 176, 186, 200
Zegris, 247
Zegry, Ez-, 253, 254
Ziryāb, 81–2
Zogoiby, 246
Zoraya, 247